INTERDISCIPLINARY APPROACHES

TO

FRESHWATER WETLANDS RESEARCH

Douglas A. Wilcox
Editor

Michigan State University Press
East Lansing, Michigan
1988

Published by Michigan State University Press but not
registered for copyright. Readers are not restricted in their
use of the contents, but recognition of the authors'
contributions should be noted with appropriate citations.

Printed in the United States of America

Michigan State University Press
East Lansing, Michigan 48823-5202

The paper used in this publication meets the minimum
requirements of American National Standard for Information
Sciences—Permanence of Paper for Printed Library Materials
ANSI Z39.48-1984

Library of Congress Cataloging-in-Publication Data

Interdisciplinary approaches to freshwater wetlands research.

   Includes bibliographies and indexes.
   1. Wetland ecology—United States.   2. Wetlands—United
States I. Wilcox, Douglas A.
QH104.I57 1988   574.5'26325'0973   88-42567
ISBN 0-87013-257-1

INTERDISCIPLINARY APPROACHES TO FRESHWATER WETLANDS RESEARCH

TABLE OF CONTENTS

List of Contributors	iv
Foreword	v
The Necessity of Interdisciplinary Research in Wetland Ecology: the Cowles Bog Example Douglas A. Wilcox	1
Fossil Pollen and Charcoal Analysis in Wetland Development Studies at Indiana Dunes National Lakeshore Richard P. Futyma	11
Sedimentology and Stratigraphy as Tools in Interpreting the Evolution of Wetland Areas in the Indiana Dunes National Lakeshore Todd A. Thompson	25
Effects of Ground Water on the Hydrochemistry of Wetlands at Indiana Dunes, Northwest Indiana Robert J. Shedlock, Nancy J. Loiacono, and Thomas E. Imbrigiotta	37
Effects of Coal Fly-Ash Disposal on Water Chemistry in an Intradunal Wetland at Indiana Dunes Douglas A. Wilcox and Mark A. Hardy	57
Prehistoric and Historic Trends in Acidity of Kettle Ponds in the Cape Cod National Seashore: Implications for Management Marjorie Green Winkler	69
Implications of Wetland Seed Bank Research: A Review of Great Basin and Prairie Marsh Studies Roger L. Pederson and Loren M. Smith	81
Surface Hydrology and Plant Communities of Corkscrew Swamp Michael J. Duever	97
Tolerance of Five Hardwood Species to Flooding Regimes L. H. Gunderson, J. R. Stenberg, and A. K. Herndon	119
Effects of Regulated Lake Levels on the Aquatic Ecosystem of Voyageurs National Park L. W. Kallemeyn, M. H. Reiser, D. W. Smith, and J. M. Thurber	133
Effects of Simulated Muskrat Grazing on Emergent Vegetation Thomas R. McCabe and Michael L. Wolfe	147
Index	159

LIST OF CONTRIBUTORS

Michael J. Duever, National Audubon Society, Rt. 6, Box 1877, Naples, Florida 33964

Richard P. Futyma, Biological Survey, New York State Museum, The State Education Department, Albany, New York 12230

L. H. Gunderson, South Florida Research Center, Everglades National Park, P. O. Box 279, Homestead, Florida 33030

Mark A. Hardy, U. S. Geological Survey, 810 Bear Tavern Road, West Trenton, New Jersey 08628

A. K. Herndon, South Florida Research Center, Everglades National Park, P. O. Box 279, Homestead, Florida 33030

Thomas E. Imbrigiotta, U. S. Geological Survey, 810 Bear Tavern Road, West Trenton, New Jersey 08628

L. W. Kallemeyn, Voyageurs National Park, P. O. Box 50, International Falls, Minnesota 56649

Nancy J. Loiacono, U. S. Geological Survey, 5957 Lakeside Blvd., Indianapolis, Indiana 46278

Thomas R. McCabe, U. S. Fish and Wildlife Service, 101 Twelfth Avenue, Box 20, Fairbanks, Alaska 99701-6267

Roger L. Pederson, Delta Waterfowl and Wetlands Research Station, R. R. 1, Portage la Prairie, Manitoba, Canada R1N 3A1

M. H. Reiser, Department of Biological Sciences, Northern Arizona University, Box 5640, Flagstaff, Arizona 86011

Robert J. Shedlock, U. S. Geological Survey, 208 Carroll Building, 8600 LaSalle Road, Towson, Maryland 21204

D. W. Smith, Department of Biological Sciences, Michigan Technological University, Houghton, Michigan 49931

Loren M. Smith, Department of Range and Wildlife Management, Texas Tech University, Box 4169, Lubbock, Texas 79409

J. R. Stenberg, South Florida Research Center, Everglades National Park, P. O. Box 279, Homestead, Florida 33030

Todd A. Thompson, Indiana Geological Survey, 611 N. Walnut Grove, Bloomington, Indiana 47405

J. M. Thurber, Department of Biological Sciences, Michigan Technological University, Houghton, Michigan 49931

Douglas A. Wilcox[1], National Park Service, Indiana Dunes National Lakeshore, 1100 N. Mineral Springs Road, Porter, Indiana 46304

Marjorie Green Winkler, Center for Climatic Research, Institute for Environmental Sciences, University of Wisconsin-Madison, 1225 West Dayton Street, Madison, Wisconsin 53706

Michael L. Wolfe, Department of Fisheries and Wildlife Sciences, Utah State University, Logan, Utah 84322

[1]Present address: U. S. Fish and Wildlife Service, National Fisheries Research Center-Great Lakes, 1451 Green Road, Ann Arbor, Michigan 48105

**FOREWORD: VOLUME 7 OF A SERIES RESULTING FROM THE
1986 CONFERENCE ON SCIENCE IN THE NATIONAL PARKS**

The 1986 Conference on Science in the National Parks was the fourth in a series of conferences conducted by the United States National Park Service (NPS) and the George Wright Society to discuss the role that science should play in supporting the understanding, management, and preservation of park resources. The dialogue focused upon the special relationship between research and management that must succeed if appropriate information is to be available to the National Park Service decision process. The Conference highlighted the active regional science programs and displayed the results of many effective research and resources management projects.

Many persons contributed their expertise and thoughts to the success of the Conference. There were over 400 attendees who gave approximately 325 poster presentations in 28 symposia. There were two plenary research panel discussions on the role of research and 12 plenary presentations on topics of importance to NPS researchers and resource managers. The Conference was attended by NPS Directorate, superintendents, researchers (natural, cultural, and social sciences), resource managers, interpreters, representatives of university and other agency research, park service personnel of six foreign countries, and the public.

The refereed papers in this volume were specifically invited by Dr. Wilcox so that they covered many of the interactions between the various disciplines used in wetlands research. It is only through awareness of the complexity of these interactions that wetlands research can be designed and conducted to accurately address the informational needs of land managers.

Raymond Herrmann,
Conference Organizer
National Park Service
Colorado State University
Ft. Collins, Colorado

THE NECESSITY OF INTERDISCIPLINARY RESEARCH IN WETLAND ECOLOGY: THE COWLES BOG EXAMPLE

Douglas A. Wilcox[1,2]

ABSTRACT

The importance of incorporating results from a number of scientific disciplines into the interpretation of wetland functions and processes was assessed by reviewing the history of research conducted in the Cowles Bog Wetland Complex in northwest Indiana. Early twentieth century work consisted primarily of descriptive studies that provided a historical reference for later work. The major research effort in the wetland was in direct response to hydrologic disturbances associated with industrial development adjacent to the site in the 1960s and 1970s. Geohydrology, surface-water hydrology, water chemistry, soil chemistry, stratigraphy, plant ecology, and animal ecology studies were all initiated at that time. These studies were continued after the industrial threats had lessened in an effort to better understand the wetland and ensure wise management of its resources. The studies also provided a framework for research on the developmental history of the wetland and its vegetation. Paleoecology, sedimentology, and remote sensing studies were added to the overall research effort to help delineate that history. The many disciplines used in the study of Cowles Bog were interrelated, and each provided information necessary for accurate interpretation of results from other studies.

INTRODUCTION

Wetland ecology encompasses a variety of scientific disciplines. Research practitioners may focus on studies of plant or animal communities, selected taxa, endangered species, hydrology, soil or water chemistry, geologic setting, or specialties such as paleoecology, seed bank analyses, or remote sensing. These studies may result in significant gains in knowledge within their own specialized fields, but they will not result in a reasonable understanding of the many functions and processes of wetlands unless they are interpreted together. Unfortunately, time and monetary constraints usually do not allow the luxury of possessing all of the interdisciplinary data necessary to draw completely accurate conclusions. As a result, management recommendations and decisions must often be made with a far-from-complete data set, and research and management personnel may lose sight of the value of gathering data from multiple disciplines.

---

[1] National Park Service, Indiana Dunes National Lakeshore, 1100 N. Mineral Springs Rd., Porter, Indiana 46304
[2] Present address: U. S. Fish and Wildlife Service, National Fisheries Research Center-Great Lakes, 1451 Green Road, Ann Arbor, Michigan 48105

The purpose of this paper is to illustrate the value of an interdisciplinary data set by presenting a history of the research conducted in the Cowles Bog Wetland Complex (CBWC). The site has historical significance, a recent, controversial political significance, and a strong record of scientific research in a number of disciplines that makes it one of the more well-studied peatlands in the Great Lakes region.

STUDY SITE

The Cowles Bog Wetland Complex is a mixture of wetland and peatland communities that occupies approximately 80 ha of the basin between the Calumet and Tolleston dunes on the south shore of Lake Michigan in Porter County, Indiana (Reshkin, 1981). This interdunal wetland was included within the boundaries of Indiana Dunes National Lakeshore by the 1966 park-authorizing legislation, and a 22 ha peatland within the wetland was designated a National Natural Landmark in that same year. Cowles Bog is more properly termed a "rich fen" (Sjörs, 1950) or "spring mire" (Moore and Bellamy, 1974) because its major source of water is a highly mineralized, artesian flow of ground water (Wilcox et al., 1986). The upwelling of ground water caused a peat mound to form in the wetland, and the local hydrologic conditions have been correlated to the development of the vegetation types within the wetland (Wilcox et al., 1986). The vegetation types are: Typha marsh, Phragmites/Typha marsh, Carex/Calamagrostis marsh, shrub swamp, Larix laricina swamp, Thuja occidentalis swamp, and Acer rubrum swamp (Figure 1).

HISTORY OF RESEARCH IN THE COWLES BOG WETLAND COMPLEX

Early Studies

Professor Henry Chandler Cowles of the University of Chicago, often called the father of plant ecology in North America, led his classes as well as the 1913 International Phytogeographic Excursion (Tansley, 1913) to the site that was then called Cowles Tamarack Swamp (Brennan, 1923; Lyon, 1927; Pepoon, 1927). Herman Kurz, one of Cowles' graduate students, conducted research on the relationships between vegetation and pH in the wetland (Kurz, 1923, 1928) and was the first to use the term "Cowles Bog" in his published studies. Early descriptions of the wetland were published by Tansley (1913), Bailey (1917), Brennan (1923), Lyon (1927), and Cressey (1928). During that time period, much of the non-forested wetland was dominated by a sedge-grass association, and cattails (Typha) were not widespread. Grass-shrub-herb associations and shrub associations graded to swamp forests containing tamarack (Larix laricina), red maple (Acer rubrum), white pine (Pinus strobus), yellow birch (Betula lutea), and northern white cedar (Thuja occidentalis). The early descriptions of Cowles Bog provided a reference from which to assess changes that would result from continued development of the region by man.

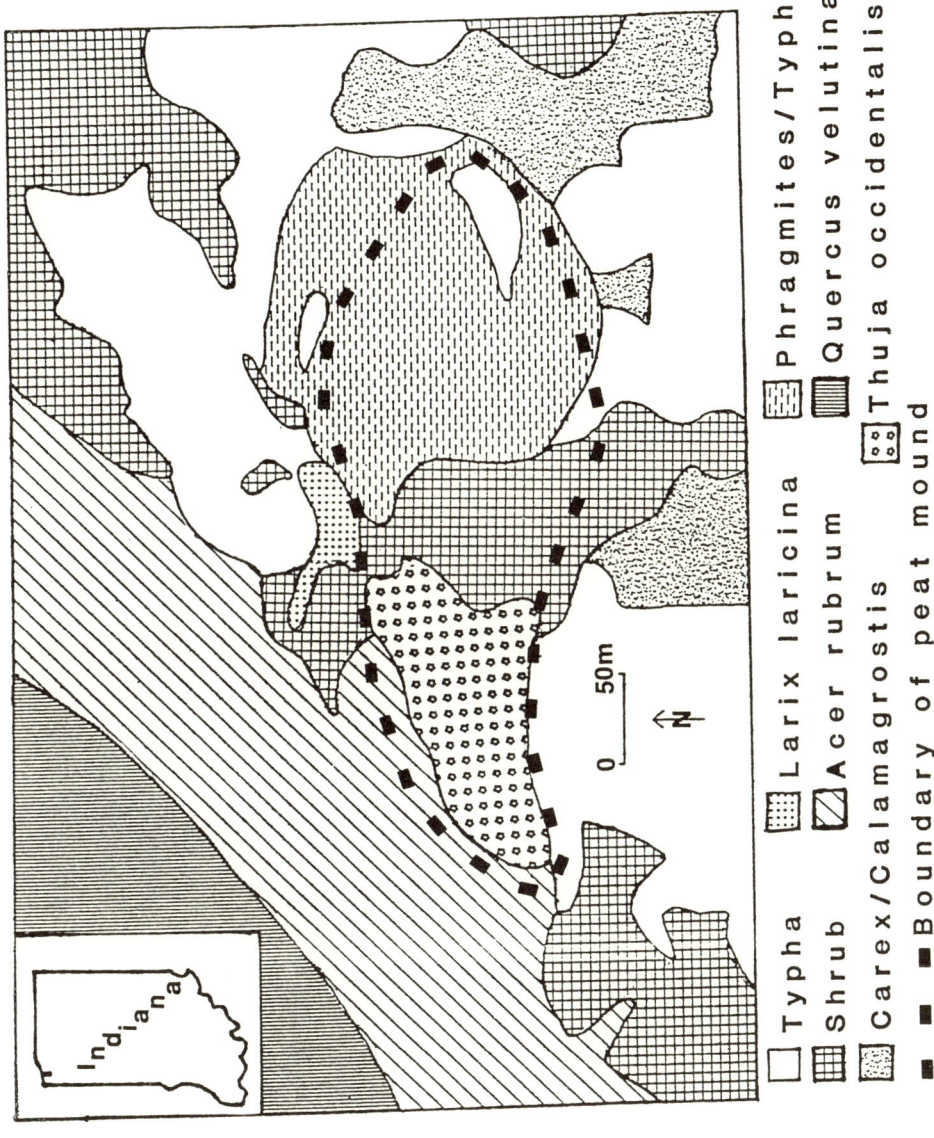

Figure 1. Map of Cowles Bog National Natural Landmark (adapted from Wilcox et al., 1986).

Environmental Impact Studies

Following the early activity, very little work was done in the CBWC until the mid-1970s when the wetland was threatened by actions associated with nearby industrial development. Specifically, a coal-fired generating station was constructed about 2 km west of the wetland in the mid 1960s, and fly-ash settling ponds were constructed within 0.7 km of the wetland. Still other ponds were constructed adjacent to the CBWC as future storage sites for fly ash dredged from the settling ponds. Seepage of water through the sand dikes of the ponds raised the water levels in adjacent wetlands of the park and also posed the threat of contamination from chemical constituents that leached from the fly ash.

In the mid-1970s, construction began on a nuclear power plant at the same industrial site. The excavation for the building foundation was dug deeper than the ground-water table at the site, so dewatering or pumping of water from the excavation was required to keep it dry. This action would create a local decline in the water table and posed the threat of lowering ground-water levels in the park. Neither action, flooding nor dewatering, was acceptable to the National Park Service (NPS), an agency charged with preserving the natural character of the land. Studies were therefore initiated to determine the magnitude of the actions on water levels in the park and to learn more about the wetlands so that the impacts of water-level changes could be assessed.

The initial efforts to study water level changes were contracted out by the power company and involved monitoring of surface-water levels in the fly-ash ponds, intradunal ponds, and the CBWC, ground-water levels in a small number of observation wells, and daily precipitation records (Texas Instruments Ecological Services, 1974-1981). The U.S. Geological Survey (USGS) was contracted by the NPS to develop a computer simulation model that would predict the amount of water-level change due to the dewatering activities. The boundaries of the two-dimensional model of the surface aquifer that was developed (Marie, 1976) included the western portion of the CBWC, but the study did not provide very much information related directly to the wetland.

The initial studies to gather baseline ecological data included biological surveys by Texas Instruments Ecological Services (1974-1981) for vegetation, mammals, birds, reptiles, amphibians, and invertebrates. A USGS study of water chemistry in the park (Arihood, 1975) reported values from the CBWC that would classify the peatland as a rich fen (specific conductance, 387 μS/cm; pH, 7.4; alkalinity, 176 mg/L as $CaCO_3$; calcium, 43 mg/L; magnesium, 19 mg/L). Classification as a fen was formally given by Carter and Stottlemyer (1978) in their ecological assessment of the wetland. They established 3 transects across the CBWC and described vegetation, soil materials, and general hydrologic conditions in an attempt to predict the effects of water-level changes on the wetland vegetation.

Boelter (1978) conducted somewhat more-detailed stratigraphic studies along one of the transects established by Carter and Stottlemyer. He assessed the types of peat materials encountered and made note of the presence of shell layers and marl deposits. He reported the presence of a peat mound in the wetland and speculated on the developmental history of the entire wetland basin, a process that involved major changes in water levels through time. Boelter also classified Cowles Bog as a fen and pointed out the importance of ground-water flow to the system. Soil chemistry analyses of samples collected along 2 of the Carter and Stottlemyer transects showed elevated levels of boron, iron, strontium, zinc, aluminum, and sulfur that were speculated to be of industrial origin (Patterson and Fenn, 1978).

A more-sophisticated, three-dimensional model for predicting water-level changes associated with construction dewatering was developed by the USGS (Meyer and Tucci, 1979) in response to changes in the dewatering plans. The new model also simulated the effects of halting seepage from the fly-ash settling ponds (in response to an agreement between the Department of the Interior and the power company to seal the ponds). The entire CBWC was included within the boundaries of the new model, which considered flow in 2 main aquifer units--the surficial water-table aquifer and a lower confined aquifer. The modeling study utilized water-level measurements from a series of observation wells around the CBWC area and generated considerable hydrologic data. The model predicted that water levels in the surface aquifer at Cowles Bog could decline by as much as 1 foot (30 cm) when the fly-ash settling ponds were sealed, and that construction dewatering for the nuclear power plant would have little effect on the Cowles Bog Wetland Complex.

A stratigraphic cross-section of the CBWC showing greater detail than Boelter's (1978) work was prepared by Hendrickson and Wilcox (1979). They first reported the clay layer that confines the deeper aquifer in Cowles Bog and also reported that wells screened in sands beneath the peat mound flowed at the peat surface. The flowing wells indicated that there was a discontinuity in the clay confining-layer beneath the mound. More extensive collections of water chemistry data utilized boron concentrations to trace waters from the fly-ash settling ponds into the southern part of the CBWC. Water chemistry data also confirmed the classification of the peatland as a rich fen (mean specific conductance, 844 µS/cm; mean pH, 7.14; mean alkalinity, 404 mg/L as $CaCO_3$; mean calcium, 88.6 mg/L; mean magnesium, 52.6 mg/L). The vegetation was found to be characteristic of a fen, also.

Further changes in the construction dewatering plans were made in 1979 that included an artificial ground-water recharge system in the surface aquifer and pumping of water from the confined aquifer beneath the construction site. These changes, coupled with the knowledge that the confined aquifer was breached beneath the peat mound of Cowles Bog, prompted further USGS computer-model simulations to determine the effects of dewatering on the CBWC (Gillies and Lapham, 1980). The new simulations predicted that water levels in the surface aquifer could decline by as much as 0.7 feet (21 cm) at the Cowles Bog peat mound under the new plans.

The final chapters on the direct impacts of power plant operations on the CBWC were studies by Hardy (1981) on the effects of fly-ash seepage on water quality and by Cohen and Shedlock (1986) on the changes in water levels and water quality following termination of the seepage in 1980. Construction activities at the nuclear plant site were terminated in 1981 by cancellation of the project.

Recent Ecological Studies

The modern floristic composition of the CBWC was reported by Wilhelm (1980) and surveys for threatened and endangered plant species were conducted by Bowles et al. (1984, 1986). Further surveys of vegetation, small mammals, and birds were conducted by Apfelbaum et al. (1983), and Titlow (1986) mapped the vegetation types in the CBWC through photointerpretation, according to the classification scheme developed and used for the entire park.

Remote sensing was also utilized by Wilcox et al. (1984) to analyze vegetation changes in the CBWC through time. Four major vegetation types were mapped from black/white and color aerial photographs in each of 9 years between 1938 and 1982. The results, replacement of large areas of sedge/grass meadow by cattail marsh, were compared to the hydrologic disturbance history of the wetland. Elevated and stabilized water levels in the CBWC, associated with the construction of one of the diked ponds by the power company, were concluded to be the main factors contributing to the invasion of cattails.

Additional stratigraphic, hydrologic, water chemistry, and vegetation studies in Cowles Bog were conducted in the early 1980s through the cooperative efforts of the NPS and the USGS. Detailed topographic and stratigraphic maps of the wetland further defined the nature of the peat mound, the artesian water supply was explained by the stratigraphy, water chemistry differences were explained by hydrologic data, plant species composition was related to water chemistry, and vegetation patterns were related to water levels/hydrology/topography. The information obtained in these studies was also correlated to changes in Lake Michigan water levels over the past 6000 years, and a reasonably-informed hypothesis of the sequence of events that led to formation of the wetland was developed (Wilcox et al., 1986). Still further studies of regional geology and hydrology formed the basis for understanding the regional ground-water flow system that causes the artesian flow of water to Cowles Bog (Shedlock, Wilcox, and Thompson, in review).

Other studies have provided great insight into the developmental history of the CBWC. Sedimentology studies by Thompson

(1986) gave many details of the geological processes involved, as well as very detailed stratigraphic information. Fossil pollen and charcoal analysis of a peat core from Cowles Bog by Futyma (1985) explained the changes in vegetation that accompanied each developmental stage of the wetland and corroborated changes in water levels that were hypothesized from stratigraphic data. Fossil mollusc analyses (Miller and Thompson, 1987) from the sediment cores of Thompson provided new faunal data and further corroborated the water-level changes hypothesized.

Current research efforts in Cowles Bog include a continuation of fossil mollusc studies (Miller, unpublished) and an assessment of airborne heavy metal deposition related to regional industrial development, as recorded in peat soil profiles (Cole, unpublished).

CONCLUSIONS

Research has been conducted in the Cowles Bog Wetland Complex utilizing a number of scientific disciplines, including geohydrology, surface water hydrology, stratigraphy/sedimentology, water chemistry, soil chemistry, paleoecology, remote sensing, plant ecology, and animal ecology. Each study could have been used solely to add new information to the data banks of its own discipline or to make limited recommendations for management of the wetland. Instead, the interdisciplinary data set has been used to foster a broad understanding of the forces at work in Cowles Bog and of the developmental history of the basin. With this knowledge, management decisions can be made with much less chance for error.

The disciplines involved also have many common denominators, and results from one study may not be readily interpretable without results from another (Wilcox, 1987). In the Cowles Bog research, studies of stratigraphy and sedimentology related directly to an understanding of ground-water hydrology, which in turn affected surface-water hydrology. Water chemistry and soil chemistry were influenced by hydrology, and all of these factors exerted strong influences on biotic development and structure within the wetland. Remote sensing, field studies of plant and animal communities, and analyses of sediment cores for pollen and macrofossil records provided modern and historic information about wetland biota. The paleoecology studies, coupled with radiocarbon dating, related directly back to the stratigraphic studies. Clearly, there was great value in working from a data set that included results from many scientific disciplines. Within the contraints of time and money, such an approach should be the goal of most wetland research efforts.

LITERATURE CITED

Apfelbaum, S. I., K. Heiman, and J. Probes, 1983. Ecological condition and management opportunity for the Great Marshsystem, Indiana Dunes National Lakeshore. Report to Indiana Dunes National Lakeshore. 60 pp.

Arihood, L. D., 1975. Water-quality assessment of the Indiana Dunes National Lakeshore, 1973-74. U.S. Geological Survey Water-Resources Investigations Report 14-75. 56 pp.

Bailey, E. S., 1917. The Sand Dunes of Indiana. McClurg, Chicago.

Boelter, D. H., 1978. An appraisal of the organic deposits in the interdunal wetlands of the Indiana Dunes National Lakeshore. Special Report USDA-Forest Service, Washington. 15 pp.

Bowles, M. L., W. J. Hess, and M. M. DeMauro, 1984. An Assessment of the Monitoring Program for Special Floristic Elements at Indiana Dunes National Lakeshore. Phase I. The Endangered Species. Report to Indiana Dunes National Lakeshore. 333 pp.

Bowles, M. L., W. J. Hess, and M. M. DeMauro, 1986. An Assessment of the Monitoring Program for Special Floristic Elements at Indiana Dunes National Lakeshore. Phase II. Threatened and Special Concern Species. Report to Indiana Dunes National Lakeshore. 375 pp.

Brennan, G. A., 1923. The Wonders of the Dunes. Bobbs-Merrill, Indianapolis. 326 pp.

Carter, V. and R. Stottlemyer, 1978. Ecology of Cowles Bog Wetland Complex. Indiana Dunes National Lakeshore Special Study. National Park Service, Porter, Indiana. 38 pp.

Cohen, D. A. and R. J. Shedlock, 1986. Shallow ground-water flow, water levels, and quality of water, 1980-1984, Cowles Unit, Indiana Dunes National Lakeshore. U.S. Geological Survey Water-Resources Investigations Report 85-4340. 25 pp.

Cressey, G. G., 1928. The Indiana Sand Dunes and Shore Lines of the Lake Michigan Basin. University of Chicago Press, for the Geographic Society of Chicago, Chicago. 77 pp.

Futyma, R. P., 1985. Paleobotanical studies at Indiana Dunes National Lakeshore. National Park Service, Porter, Indiana, USA. 242 pp.

Gillies, D. C. and W. W. Lapham, 1980. Reassessment of the effects of construction dewatering on ground water levels in the Cowles Unit, Indiana Dunes National Lakeshore, Indiana. Supplement to Geological Survey Water-Resources Investigations 78-138. U.S. Geological Survey Open-File Report 80-1105. 50 pp.

Hardy, M. A., 1981. Effects of coal fly-ash disposal on water quality in and around the Indiana Dunes National Lakeshore, Indiana. U.S. Geological Survey Water-Resources Investigations Report 81-16. 64 pp.

Hendrickson, W. H. and D. A. Wilcox, 1979. Relationship between some physical properties and the vegetation found in Cowles Bog National Landmark, Indiana. Proceedings of Second Conference on Scientific Research on the National Parks 5: 642-666.

Kurz, H., 1923. Hydrogen ion concentration in relation to ecological factors. Botanical Gazette 76: 1-29.

Kurz, H., 1928. Influence of sphagnum and other mosses on bog reactions. Ecology 9: 56-69.

Lyon, M. W., 1927. List of flowering plants and ferns in the Dunes State Park and vicinity, Porter County, Indiana. American Midland Naturalist 10: 245-295.

Marie, J. R., 1976. Model analysis of effects on water levels at Indiana Dunes National Lakeshore caused by construction dewatering. U.S. Geological Survey Water-Resources Investigations Report 76-82. 32 pp.

Meyer, W. and P. Tucci, 1978. Effects of seepage from fly-ash settling ponds and construction dewatering on ground-water levels in the Cowles Unit, Indiana Dunes National Lakeshore, Indiana. U.S. Geological Survey Water-Resources Investigations Report 78-138. 95 pp.

Miller, B. B. and T. A. Thompson, 1987. Molluscan faunal changes in the Cowles Bog area (Indiana Dunes National Lakeshore) following the low-water Lake Chippewa phase. In: Schneider, A. F. and G. S. Fraser (eds.) Geological Society of America Special Paper (in press).

Moore, P. D. and D. J. Bellamy, 1974. Peatlands. Springer-Verlag, New York. 221 pp.

Patterson, J. C. and D. G. Fenn, 1978. A report on the analyses of selected soil substrates from Indiana Dunes National Lakeshore. Indiana Dunes National Lakeshore Special Study. National Park Service, Porter, Indiana. 88 pp.

Pepoon, H. S., 1927. An Annotated Flora of the Chicago Area. Chicago Academy of Sciences. 554 pp.

Reshkin, M., 1981. Geology and Soils of the Indiana Dunes National Lakeshore. Vol. I, Indiana Dunes Research Program Report 81-01. 144 pp.

Sjörs, H., 1950. On the relation between vegetation and electrolytes in northern Swedish mire waters. Oikos 2: 241-258.

Tansley, A. G., 1913. International Phytogeographic Excursion (I.P.E.) in America, 1913. New Phytologist 12: 322-336.

Texas Instruments Ecological Services, Inc., 1974-1981. Annual Reports. Bailly Nuclear-1 Site. Northern Indiana Public Service Company. Texas Instruments, Inc., Dallas.

Thompson, T. A., 1986. Sedimentology, internal architechture, and depositional history of the Indiana Dunes National Lakeshore and State Park. Ph.D. Thesis. Indiana University, Bloomington, Indiana.

Titlow, B., 1986. Vegetation mapping of Indiana Dunes National Lakeshore. Indiana Dunes National Lakeshore special study. National Park Service, Porter, Indiana.

Wilcox, D. A., 1987. A model for assessing interdisciplinary approaches to wetland research. Wetlands 7: 39-50.

Wilcox, D. A., S. I. Apfelbaum, and R. D. Hiebert, 1984. Cattail invasion of sedge meadows following hydrologic disturbance in the Cowles Bog Wetland Complex, Indiana Dunes National Lakeshore. Wetlands 4: 115-128.

Wilcox, D. A., R. J. Shedlock, and W. H. Hendrickson, 1986. Hydrology, water chemistry, and ecological relations in the raised mound of Cowles Bog. Journal of Ecology 74: 1103-1117.

Wilhelm, G. S., 1980. Report on the special vegetation of the Indiana Dunes National Lakeshore. INDU Research Program Report 80-01. 262 pp.

FOSSIL POLLEN AND CHARCOAL ANALYSIS IN WETLAND DEVELOPMENT STUDIES AT INDIANA DUNES NATIONAL LAKESHORE

Richard P. Futyma[1]

ABSTRACT

The techniques of fossil pollen and charcoal analysis, along with radiocarbon dating, were used to elucidate the histories of three wetland areas in Indiana Dunes National Lakeshore: Pinhook Bog, Cowles Bog, and the Miller Woods Ponds. At both Pinhook Bog and Cowles Bog the aquatic pollen records begin with lake floras of floating-leaved aquatics (water-lily, yellow pond-lily, and water-shield). The lake phase at Pinhook Bog began prior to 12,000 years Before Present; starting ca. 8000 yr B.P., a floating mat of Sphagnum moss, sedges, heath shrubs, and conifers (white pine and larch) expanded from the edge of the lake, covering most of its area by 2000 yr B.P. At Cowles Bog, the lake phase began ca. 7000 yr B.P. and was replaced about 2000 yr B.P. by a grass-dominated marsh that was soon colonized by Sphagnum, larch, and white pine. During the last 100 years, the conifer swamp was transformed to one dominated by hardwoods as a result of human activities. The Miller Woods ponds were formed within the last 4000 years and have aquatic floras similar to those of the lake phases of Cowles and Pinhook bogs. Recent human disturbance has led to extensive invasion of cattails in some ponds, greatly changing their character.

INTRODUCTION

Park managers are frequently faced with the task of preserving habitats or restoring them to their "natural" condition. A first step in this process, determining what the habitats were like before human impacts, is difficult because of the lack of written records describing pre-settlement plant communities. Similarly, it may be difficult to determine whether recently observed changes in plant communities are human-induced or are part of long-term, natural successional trends. Understanding of long-term changes in wetland plant communities is facilitated by studies of plant remains preserved in sediments and peat. Pollen grains are among the most abundant and best preserved of such remains and can be easily separated from the sedimentary matrix, examined, and identified. Charcoal ash particles found with the pollen provide evidence of past occurrence of fires within and near the wetland. In combination with radiocarbon dating, these fossil records may be used to reconstruct detailed histories of plant communities.

At Indiana Dunes National Lakeshore, on the southern shore of Lake Michigan, these techniques were used to study three wetland areas: Pinhook Bog, Cowles Bog, and the interdunal ponds of Miller Woods. The primary questions behind the studies were the manner

---
[1]Biological Survey, New York State Museum, Albany, New York 12230

in which the wetlands formed and the degree to which these communities have been altered by humans.

METHODS

Wetland sediments used for pollen and charcoal analyses were collected by means of Livingstone-, Hiller-, Davis-, and Jowsey-type coring devices (Moore and Webb, 1978). Samples a few cubic centimeters in volume taken from various levels in the cores were treated to isolate pollen and charcoal through mechanical means, as by sieving, and by treatment with chemical reagents to remove mineral particles and unwanted organic matter (Faegri and Iversen, 1975; Birks and Birks, 1980; Doher, 1980). The resulting pollen-rich residues were suspended in silicone fluid and mounted on microscope slides. Magnifications of 400x and greater were used in examining and counting pollen grains and charcoal particles. Fossil pollen grains were identified with the aid of pictorial keys (Kapp, 1969; McAndrews et al., 1973) and reference slides of pollen collected from living plants and herbarium specimens.

A fossil pollen stratigraphy is depicted in a pollen diagram, which consists of a series of graphs showing change in abundance of identified pollen types with depth. In the cases discussed here, pollen abundance at a given depth is expressed as a percentage of the sum of pollen from all trees and terrestrial herbs and shrubs identified at that depth. Pollen of wetland and aquatic plants are excluded from the sum, but their percentages are based on that sum; therefore, it is possible for these taxa to have abundances greater than 100%. Use of the terrestrial sum provides a relatively stable "background" pollen rain against which to compare the more irregular inputs of wetland taxa. Confidence intervals for pollen percentages were calculated by multinomial statistical methods (Mosimann, 1965). Abundance of charcoal particles in the sediment is expressed in terms of charcoal area (square micrometers) per 100 pollen grains of the terrestrial sum. As an aid in their interpretation, the pollen stratigraphies were divided into zones in which the pollen assemblages are relatively homogeneous; numerical techniques developed by Gordon and Birks (1972) were used.

The interpretation of a fossil pollen stratigraphy in terms of vegetational history is an interdisciplinary exercise. Factors that must be taken into account include the general relationship between abundance of plants and abundance of their pollen in the "pollen rain" (Delcourt et al., 1984), the relationship between the size of the sediment-collecting basin and the size of the area from which the pollen it contains was derived (Jacobson and Bradshaw, 1981), sedimentary processes affecting the way in which pollen is distributed within a basin (Davis et al., 1984), and both biotic and abiotic factors affecting the dispersal and preservation of pollen grains. In studies of wetland development, vertical variations in the characteristics of the sediments from which the pollen was obtained provide important additional clues to environmental changes. Weight losses upon ignition (L.O.I.) of dried sediment samples, burned first at 550C, then at 1000C, were measured to determine the proportions of organic matter and carbonates (Dean, 1974).

RESULTS AND DISCUSSION

Pinhook Bog

Pinhook Bog occupies an 18-m-deep, 40-ha kettle-hole on the Valparaiso moraine, 15 km south of Lake Michigan (Figure 1). It is covered by a Sphagnum moss mat with a variety of peatland plants, including Sarracenia purpurea (pitcher-plant), Drosera intermedia (sundew), Vaccinium oxycoccos (small cranberry), V. corymbosum (highbush-blueberry), V. atrococcum (black highbush-blueberry), Chamaedaphne calyculata (leather-leaf), Pyrus floribunda (chokeberry), Gaylussacia baccata (huckleberry), Nemopanthus mucronata (mountain-holly), Acer rubrum (red maple), Larix laricina (larch), and Pinus strobus (white pine). Water chemistry data indicate weakly minerotrophic conditions in the peatland (Wilcox, 1986).

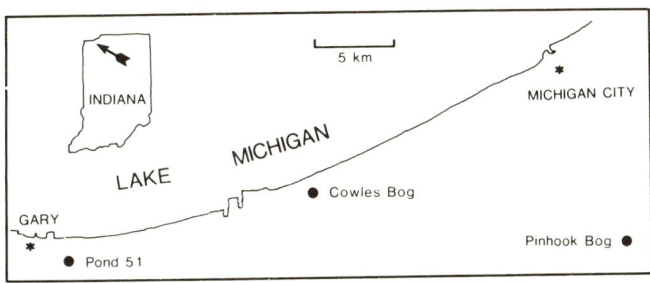

Figure 1. Map of northwestern Indiana showing Indiana Dunes National Lakeshore and the locations of Pinhook Bog, Cowles Bog, and Pond 51 of Miller Woods.

At one of the deeper portions of the basin, a 13.15-m sediment core was retrieved; its pollen stratigraphy is shown in Figure 2. From the beginning of the record, ca. 12,000 yr B. P., to about 4000 yr B.P., the basin was filled with a lake, as indicated by the presence of pollen of floating-leaved aquatics, e.g., Nuphar (yellow water lily), Nymphaea (water lily), and Brasenia (water shield), and remains of the green alga Pediastrum. About 8000 yr B.P. Cephalanthus (buttonbush) became abundant in shallow water around the lake (pollen zone PB-2A), possibly as a result of the same early Holocene climatic changes (Bartlein et al., 1984) that caused the replacement of pine-dominated forests by forests of oak and other hardwoods on the surrounding uplands. Increases in Cyperaceae pollen at the same time may indicate the beginning of a floating sedge mat on the lake's perimeter. By 5000 yr B.P., the floating mat had become large enough and close enough to the coring site that detritus dislodged from the mat contributed a sizable component to the lake sediments.

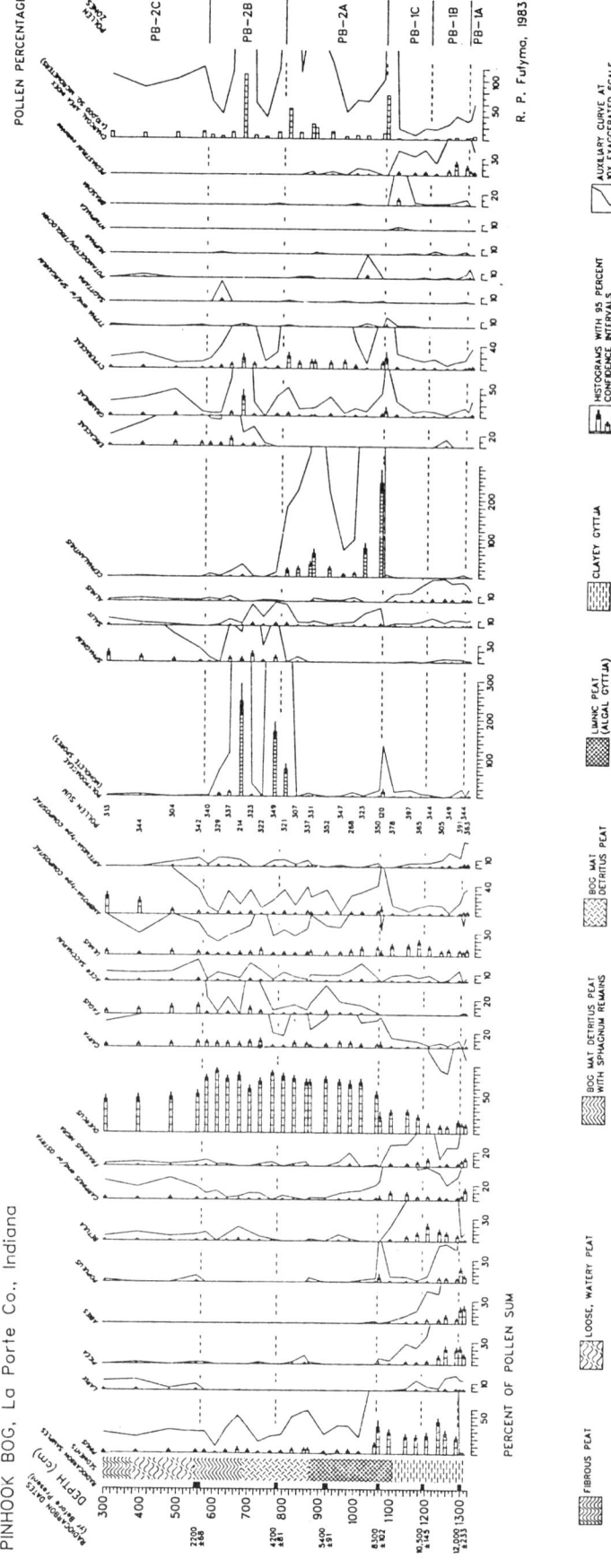

Figure 2. Fossil pollen stratigraphy of Pinhook Bog. Taxa excluded from the pollen sum are plotted to the right of the pollen sum column. The top 300 cm of peat at this location could not be sampled adequately because it was very fibrous. See appendix for laboratory identification numbers of radiocarbon dates.

Pollen zone PB-2B marks the beginning of a bog mat with *Sphagnum* moss and ericaceous shrubs; the abundant Polypodiaceous fern spores in this zone may be from ferns such as *Thelypteris palustris* (marsh fern) growing in the mat. Larch trees did not become abundant in the mat until ca. 2000 yr B.P.; the disappearance of pollen of floating-leaved aquatics and *Cephalanthus* may indicate that the lake's surface had become almost completely covered by the bog mat at that time. *Cephalanthus* is insect-pollinated, and most of its pollen incorporated into the lake deposits would have fallen directly into the water rather than having been carried by wind. An expanding mat of vegetation between the open water and the shrub zone would have decreased the amount of water-borne *Cephalanthus* pollen reaching the lake. Even today there is a large amount of *Cephalanthus* growing in the moat at the outer edge of the bog, but little of its pollen reaches the site where the core was taken, only 150 m away.

Loss-on-ignition data for the sediments illustrate the changes in the environment of deposition (Figure 3a). Limnic sediments, especially below 950 cm depth, have a large inorganic component, including silt, clay, and diatom frustules, and L.O.I. is generally less than 40%. As a floating bog mat grew out over the lake, inorganic inputs decreased and organic detritus from the mat became an important part of the sediment, raising L.O.I. above 70%. Sediments composed entirely of plant remains from the bog mat, as in samples above 560 cm, have L.O.I. values close to 95%.

The upper 2.5 m of peat at this site was too fibrous to sample easily, so another core was taken 60 m away. The combined record of the two cores shows that no major vegetational changes have occurred in Pinhook Bog in the last 2000 years (Futyma, 1985).

Cowles Bog

On the Calumet lake plain, about 0.5 km south of Lake Michigan, is the Cowles Bog Wetland Complex. It includes a fen-like area of cattails and sedges and a swamp forest with various mixtures of *Larix laricina*, *Pinus strobus*, *Thuja occidentalis* (white cedar), *Acer rubrum*, *Betula alleghaniensis* (yellow birch), *Fraxinus americana* (white ash), and *F. nigra* (black ash). As is typical of sites influenced by mineral-rich, circum-neutral ground water, *Sphagnum* moss occurs only in small, isolated clumps. In one part of the peatland, ground water upwelling under artesian pressure has caused the development of a peat mound that rises ca. 1 m above the surrounding surface (Wilcox et al., 1986).

The development of Cowles Bog (Figure 4) began with a lake phase ca. 7000 years ago. At that time, water levels in the Lake Michigan basin, which had been rising from an early low stage, reached an elevation close to the modern level (Larsen, 1985). The rise in Lake Michigan probably elevated the local water table, flooding the Cowles Bog basin. Pollen of floating-leaved aquatics and the alga *Pediastrum* are frequent in sediments deposited before

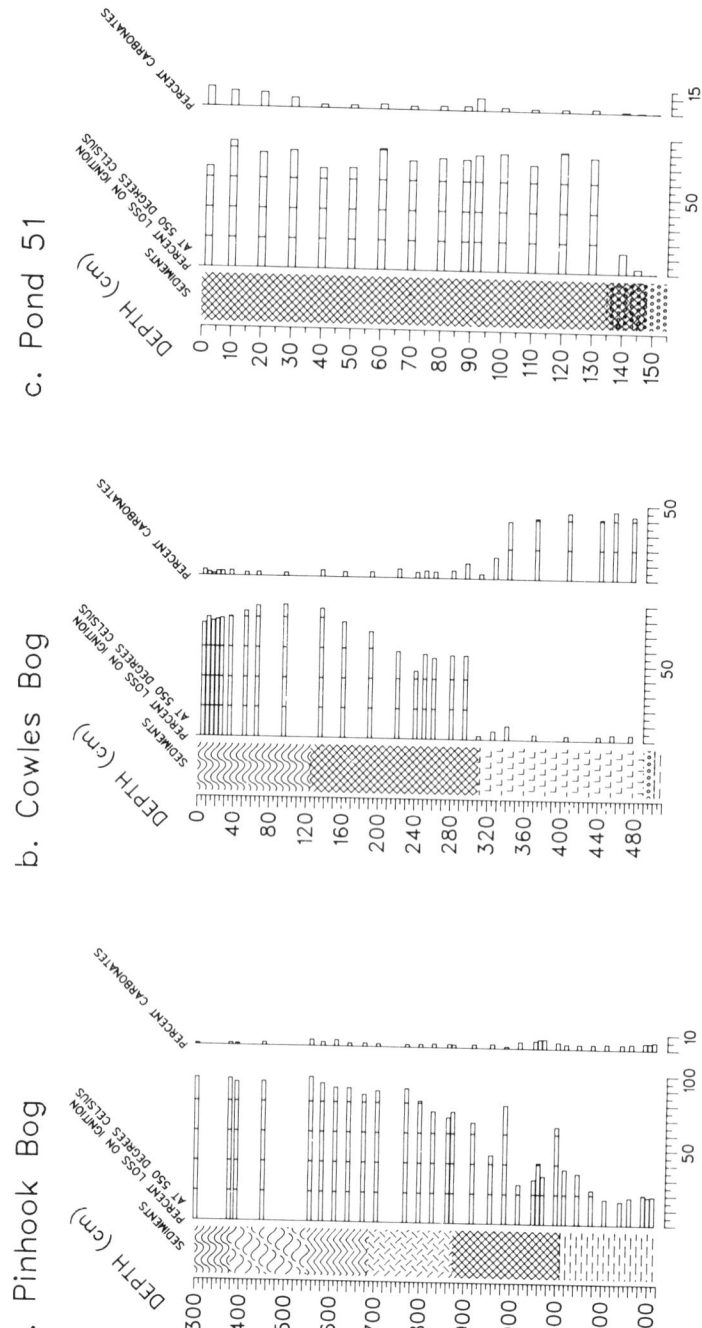

Figure 3. Loss-on-ignition and carbonate content data for the sediments of (a) Pinhook Bog, (b) Cowles Bog, and (c) Pond 51. The values are percentages of dry weight of sediment. See Figures 2, 4, and 5 for explanation of sediment symbols.

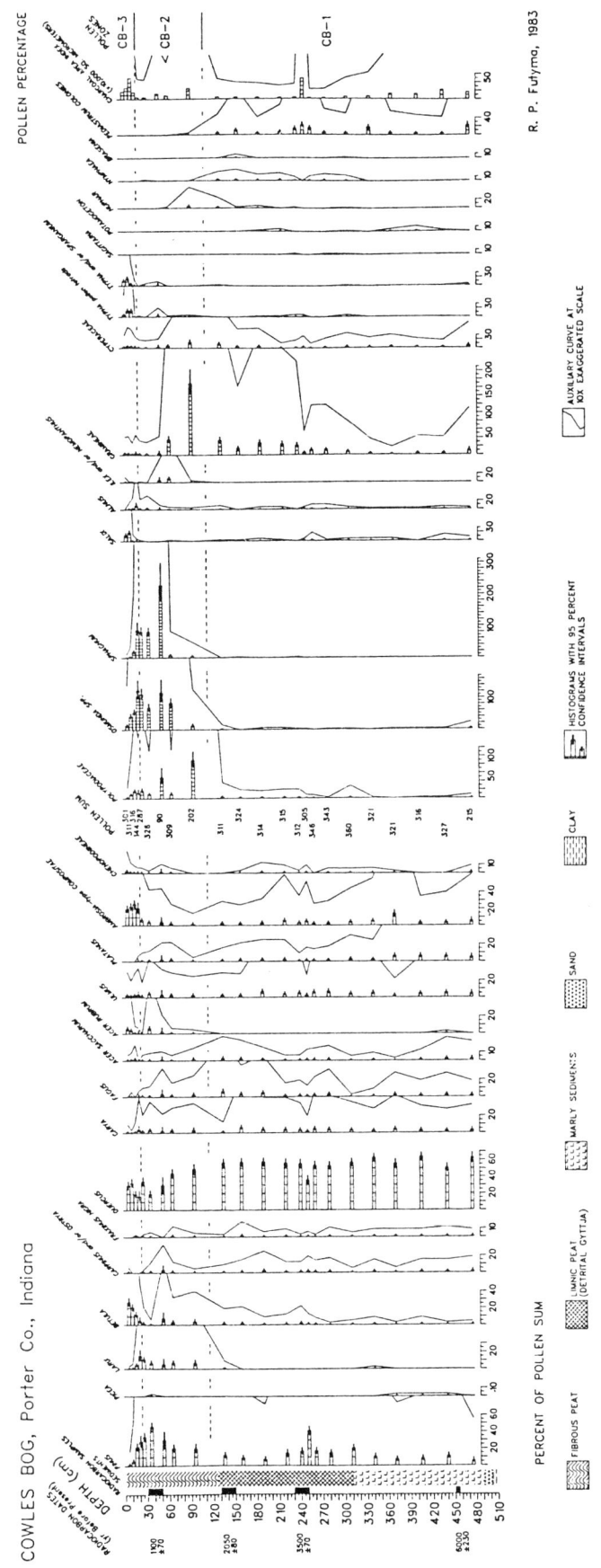

Figure 4. Fossil pollen stratigraphy of Cowles Bog. Taxa excluded from the pollen sum are plotted to the right of the pollen sum column.

1700 yr B.P. (pollen zone CB-1). The abundance of grass pollen in upper zone CB-1 may indicate occupation of shallow areas of the lake by grassy marshes, probably including some sedges. About 2000 yr B.P. (pollen zone CB-2) the site where this core was retrieved was overtaken by marsh vegetation, indicated by the the sudden rise in grass pollen and the change from limnic peat to fibrous swamp peat. This was followed shortly by a conifer swamp that included _Larix laricina_, _Pinus strobus_, _Ilex verticillata_ (winterberry) and/or _Nemopanthus mucronata_, _Alnus rugosa_ (speckled alder), Polypodiaceous ferns, _Osmunda_ spp. (royal fern), and _Sphagnum_.

Data for loss on ignition and percent carbonates (Figure 3b) reflect the relatively abrupt change from marly sediments to limnic peat. The sample from 310 cm, which had both low L.O.I. and percent carbonates, consisted mainly of non-carbonate sand. Here, as at Pinhook Bog, the peat deposited in a lake environment has moderate L.O.I. values (45-75%), whereas the fibrous peat derived from swamp vegetation has high L.O.I., generally >80%.

During the last 150 years, which is indicated by the increase in _Ambrosia_ (ragweed) pollen concomitant with European settlement of the area (zone CB-3, Figure 3), there have been dramatic decreases in pollen and spores of _Pinus_, _Larix_, Polypodiaceae, _Osmunda_, and _Sphagnum_, and significant increases in pollen of _Betula_, _Salix_, and _Typha_ (cattail). These suggest that human activities have changed the vegetation at the coring site from a conifer swamp to the hardwood swamp dominated by _Betula alleghaniensis_, _Salix nigra_ (black willow), and _Acer rubrum_ that now occupies this portion of Cowles Bog. The increase in _Typha_ pollen reflects the spread of cattails in a large, open area within several hundred meters of the coring site, which has taken place over the past 50 years (Wilcox _et al_., 1984). Attempts to drain the wetland and logging in the swamp forest are probably the main causes of these changes.

The abundance of charcoal particles in the Cowles Bog stratigraphy shows that fires have been a frequent occurrence in this region in the past 6000 years. The slight rise in charcoal abundance from zone CB-1 to CB-2, at the change from lake sediments to bog mat peat, may be a result of occasional surface fires in the peatland. A similar charcoal increase coincided with bog mat development in Pinhook Bog (zone PB-2A, Figure 2). The increase of charcoal abundance in pollen zone CB-3 is most likely due to recent activities of humans.

Miller Woods Interdunal Ponds

Miller Woods is a 2-km-wide strip along Lake Michigan encompassing a series of ponds separated by hummocky dunes and dune ridges. This ridge-and-swale topography was formed gradually over the last 4000 years by a general lowering of the level of Lake Michigan and periodic fluctuations in lake level (Olson, 1958). Younger ponds have mostly open water with few floating-leaved aquatics and sparse borders of emergent aquatics, whereas older ponds have abundant floating-leaved aquatics and are choked with cattails in shallow areas. Earlier workers (e.g., Shelford, 1911) suggested these vegetational differences between young and old ponds represent a successional sequence that has occurred or will occur in each of these ponds. This concept was questioned in recent quantitative studies by Wilcox and Simonin (1987). To test the idea, pollen stratigraphies of five of the ponds were examined. Presented here (Figure 5) is the pollen stratigraphy of Pond 51, the oldest of the ponds studied.

Pond 51 was formed about 3000 yr B.P.; its oldest sediments show no aquatic plant pollen except for a small amount of Cyperaceae. Similarly, the adjacent dunes were somewhat sparsely covered by pine-dominated vegetation. The low organic matter content of the lowermost sediments (Figure 3c) suggests that they were deposited rapidly and that the vegetation represented by pollen zone MW-1 existed only briefly. A change to oak woodland on the dunes was accompanied by colonization of the ponds by Nuphar, Nymphaea, Brasenia, Utricularia (bladderwort), Potamogeton (pondweed), Sagittaria (arrow-head), and Typha. Grasses, perhaps including Zizania aquatica (wild rice), became abundant in shallow areas and Alnus and Cephalanthus became established on the edge of the pond. Most of these taxa show little significant change in abundance of their pollen throughout the sequence (zones MW-2 and MW-3). The pollen of Typha, however, did increase significantly in the period since settlement by Europeans (indicated by increases in Ambrosia and other agricultural weeds). Therefore, the great abundance of Typha, which dominates the major portion of this pond, is probably a recent development resulting from disturbance by human activities. Also, Brasenia, which was abundant throughout most of the stratigraphy, declined in zone MW-3; it no longer is found in the pond, and its demise may also be a result of disturbance.

Overall, the pollen stratigraphy shows that for most of its history, the aquatic vegetation of Pond 51 has had a rather stable composition, undergoing little successional change. Prior to the arrival of European settlers, its vegetation was much like that now found in the younger ponds of Miller Woods. Therefore, the vegetational variation among ponds of different ages may represent varying degrees of disturbance in recent times rather than differences in successional stage.

The charcoal record for Pond 51 (Figure 5) begins with a moderately high value in zone MW-1, probably indicating inwashing of charcoal particles, along with much sand, from the dune slopes adjacent to the pond. The low but constant charcoal values in zone MW-2 may be a result of frequent fires of moderate to low intensity in the woodland around the pond. As at Cowles Bog, increases in charcoal during the period following European settlement (zone MW-3) are probably due to human activities that increased the local incidence of fires, or from the burning of wood for fuel. Therefore, the charcoal record in the sediments of

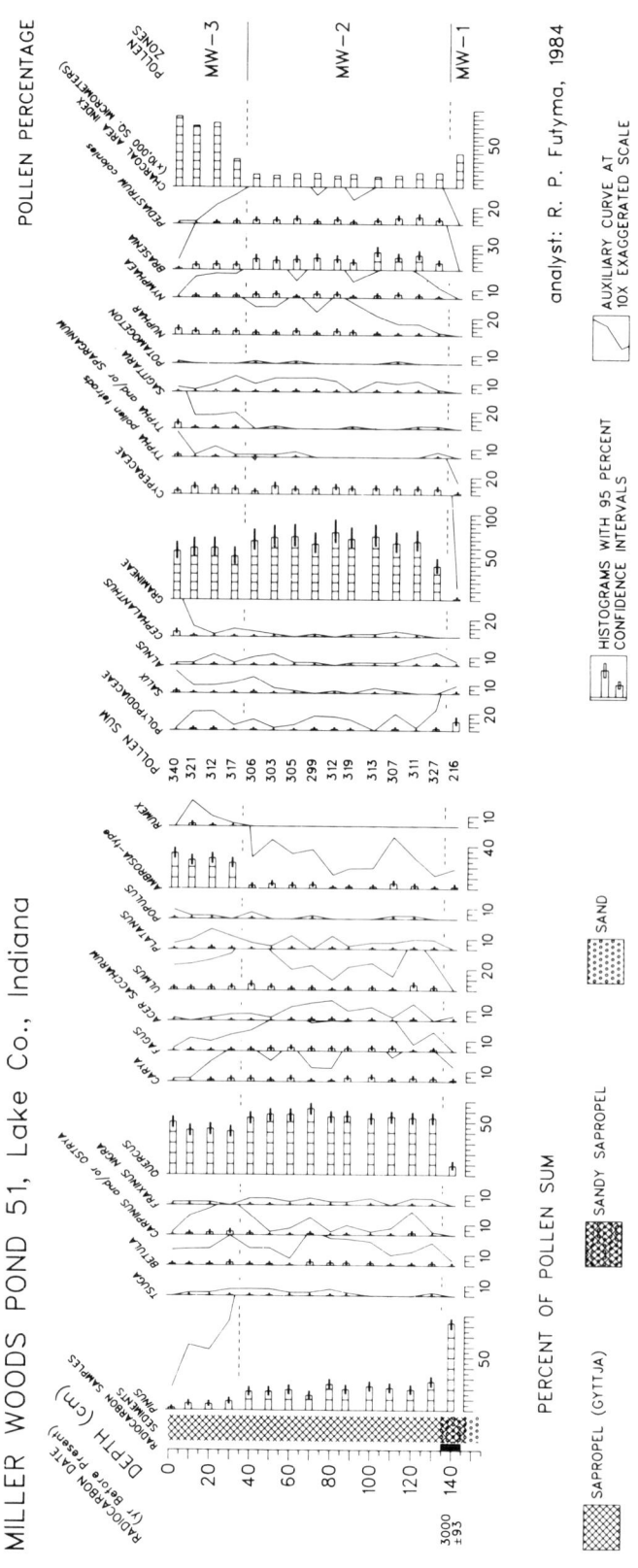

Figure 5. Fossil pollen stratigraphy of Pond 51, Miller Woods. Taxa excluded from the pollen sum are plotted to the right of the pollen sum column.

Pond 51 appears to have little bearing on the development of wetland vegetation in the basin.

CONCLUSIONS

These studies at Indiana Dunes National Lakeshore have demonstrated the utility of fossil pollen stratigraphies in expanding our understanding of wetland ecosystems, their history, and successional trends. At Pinhook Bog, the pollen record reveals the nature of the vegetation in each of the stages in the transformation of a lake to a peatland completely covered by a mat of bog vegetation. Similarly, the pollen stratigraphy of Cowles Bog documents the conversion of a lake to a conifer swamp, but also shows that recent human activities, most likely logging and attempts to drain the wetland, have greatly affected its vegetation by decreasing the abundance of conifers in the swamp and causing the spread of cattails throughout a major part of the wetland.

Detailed pollen studies in the Miller Woods ponds, described only briefly here, have provided a test of a hypothesis concerning hydrarch succession. The paleoecological perspective provided by the pollen records from this site leads us to believe that many of the changes that had been thought to be the result of natural successional processes are likely to have been recent changes due to human disturbance.

ACKNOWLEDGMENTS

This work was supported by the National Park Service through contract no. CX 6000-3-0102 awarded to The University of Michigan. D. A. Wilcox provided invaluable assistance in planning this project and during the field work. The suggestions of two anonymous reviewers helped to improve this paper. This is contribution number 510 of the New York State Science Service.

LITERATURE CITED

Bartlein, P. J., T. Webb III, and E. Fleri, 1984. Holocene climatic change in the northern Midwest: pollen-derived estimates. Quaternary Research 22:361-374.

Birks, H. J. B. and H. H. Birks, 1980. Quaternary Palaeoecology. University Park Press, Baltimore, Maryland.

Davis, M. B., R. E. Moeller, and J. Ford, 1984. Sediment focusing and pollen influx. In: Lake Sediments and Environmental History, E. Y. Haworth and J. W. G. Lund (Editors). University of Leicester Press, England, pp. 261-293.

Dean, W. E., 1974. Determination of carbonates and organic matter in calcareous sedimentary rocks by loss on ignition: comparison with other methods. Journal of Sedimentary Petrology 44: 242-248.

Delcourt, P. A., H. R. Delcourt, and T. Webb, III, 1984. Atlas of mapped distributions of dominance and modern pollen percentages for important tree taxa of eastern North America. American

Association of Stratigraphic Palynologists Contribution Series, no. 14, pp. 1-131.

Doher, L. I., 1980. Palynomorph preparation procedures currently used in the paleontology and stratigraphy laboratories, U. S. Geological Survey. U. S. Geological Survey Circular 830:1-29.

Faegri, K. and J. Iversen, 1975. Textbook of Pollen Analysis. Third edition. Hafner Press, New York.

Futyma, R. P., 1985. Paleobotanical Studies at Indiana Dunes National Lakeshore. Indiana Dunes National Lakeshore Science Division Report.

Gordon, A. D. and H. J. B. Birks, 1972. Numerical methods in Quaternary palaeoecology. I. Zonation of pollen diagrams. New Phytologist 71:961-979.

Jacobson, G. L., Jr. and R. H. W. Bradshaw, 1981. The selection of sites for paleovegetational studies. Quaternary Research 16:80-96.

Kapp, R. O., 1969. How to Know Pollen and Spores. Wm. C. Brown Co., Dubuque, Iowa.

Larsen, C. E., 1985. A stratigraphic study of beach features on the southwestern shore of Lake Michigan: new evidence of Holocene lake level fluctuations. Illinois State Geological Survey, Environmental Geology Notes 112.

McAndrews, J. H., A. A. Berti, and G. Norris, 1973. Key to the Quaternary pollen and spores of the Great Lakes region. Life Sciences Miscellaneous Publication, Royal Ontario Museum, Toronto.

Moore, P. D. and J. A. Webb, 1978. An Illustrated Guide to Pollen Analysis. Halsted Press, New York.

Mosimann, J. E., 1965. Statistical methods for the pollen analyst. Multinomial and negative multinomial techniques. In: Handbook of Paleontological Techniques, B. G. Kummel and D. M. Raup (Editors). Freeman Co., San Francisco, pp. 636-673.

Olson, J. S., 1958. Lake Michigan dune development. 3. Lake-level, beach, and dune oscillations. The Journal of Geology 66:473-483.

Shelford, V. E., 1911. Ecological succession. II. Pond fishes. Biological Bulletin 21:127-151.

Wilcox, D. A., 1986. The effects of deicing salts on water chemistry in Pinhook Bog, Indiana. Water Resources Bulletin 22: 57-65.

Wilcox, D. A., S. I. Apfelbaum, and R. D. Hiebert, 1984. Cattail invasion of sedge meadows following hydrologic disturbance in the Cowles Bog Wetland Complex, Indiana Dunes National Lakeshore. Wetlands 4:115-128.

Wilcox, D. A., R. J. Shedlock, and W. H. Hendrickson, 1986. Hydrology, water chemistry, and ecological relations in the raised mound of Cowles Bog. Journal of Ecology 74:1103-1117.

Wilcox, D. A. and H. A. Simonin, 1987. A chronosequence of aquatic macrophyte communities in dune ponds. Aquatic Botany 28: 227-242.

APPENDIX

Laboratory identification numbers for radiocarbon dates. All radiocarbon analyses were performed by the Radiochemistry Analytical Center of the U. S. Geological Survey in Denver, Colorado.

	date	lab no.
Pinhook Bog		
	2200 ±68	DE-249
	4200 ±81	DE-250
	5400 ±91	DE-251
	8300 ±102	DE-252
	10,500 ±145	DE-253
	12,000 ±233	DE-254
Cowles Bog		
	1100 ±70	DE-236
	2050 ±80	DE-237
	3500 ±70	DE-238
	6000 ±230	DE-256
Pond 51		
	3000 ±93	DE-257

SEDIMENTOLOGY AND STRATIGRAPHY AS TOOLS IN INTERPRETING THE EVOLUTION OF WETLAND AREAS IN THE INDIANA DUNES NATIONAL LAKESHORE

Todd A. Thompson[1]

ABSTRACT

Coastal wetlands are the product of a complex interaction of coastal, mainland, and intra-wetland physical and biological processes. The relative importance of these processes to the distribution of organisms and sediment types within the wetland can vary considerably through time and space. Understanding the evolution of the wetland requires an understanding of the significant processes and a knowledge of the internal architecture of the wetland. A detailed stratigraphy and internally consistent environmental interpretation provide a geologic framework on which hydrological, botanical, and ecological studies are based.

Closely-spaced vibracores through the entire wetland sequence in the Indiana Dunes National Lakeshore have defined the lateral and vertical distribution of different sediments within the Great Marsh, an extensive wetland between two dune and beach complexes, the Calumet and Toleston Beaches. The sediments and their distribution are typical of north-temperate-zone lakes, but are specific in having been influenced by a rising external lake level. That is, the level of glacial Lake Nipissing (ancestral Lake Michigan) rose as much as 20 ft (6 m) as marl and organic sediments accumulated in the back-barrier basin. Lateral changes in sediment type are related to intrabasinal drainage, predepositional topography on which the wetland developed, and degree of hydraulic connection between the wetland and underlying glacigenic sediments.

INTRODUCTION

Wetlands are common in most coastal settings, and they range from salt-water-dominated tidal marshes and mangrove forests to fresh-water-dominated bogs, fens, marshes, swamps, and lakes. Coastal wetlands form and are maintained by an interaction of physical and biological processes in the wetland and by processes imposed on the wetland by the coast and mainland. The relative importance of these processes varies from wetland to wetland but also varies across a single wetland. Moreover, most wetlands evolve, or mature (Frey and Basan, 1978). This evolution is often dominated by physical processes in the early history of the wetland basin (Weller, 1978) and becomes more biologically dominated as vegetation is established and the basin fills. The study of these formative physical processes is uniquely the realm of sedimentology and its sister field, stratigraphy.

---
[1]Indiana Geological Survey, 611 N. Walnut Grove, Bloomington, Indiana 47405

Understanding the evolution of the wetland requires a knowledge of the influence of the factors that control the type and intensity of the available process. Such a knowledge can be gained by collecting and examining a large number of cores and by developing a detailed stratigraphy and internally consistent environmental interpretation. It is on this geologic framework that hydrological, botanical, and ecological studies of the wetland can be based.

The purpose of this paper is to describe the internal architecture and development of an extensive fresh-water intradunal-wetland along the southeastern shore of Lake Michigan. Emphasis is given to the characteristics of the sedimentary deposits in the wetland basin and the major processes that governed sedimentation.

METHODS

The study area was the Great Marsh, which is entirely within the Indiana Dunes National Lakeshore and the Indiana Dunes State Park (Figure 1A). The Great Marsh is about 0.5 mi (0.8 km) wide and 11 mi (18 km) long and is bounded on the north and the south by dune and beach complexes, the Toleston Beach and Calumet Beach, respectively. Several drainages channel water through the Great Marsh. The most important are Brown, Derby, and Kintzele Ditches and Dunes Creek.

Eighty-seven vibracores were collected in the study area during the late spring and summer of 1984 and 1985 (Figure 1A). The vibracorer used in this study followed the design of Finklestein and Prins (1981). A vibracorer is an excellent tool for wetland studies because it is portable, requires only 2 people to operate, and collects an undisturbed 30-ft (9.1-m) core. My crew and I have carried the corer through many miles (kilometers) of wetland with few mishaps and always excellent results.

Most of the cores were collected in 10 linear transects perpendicular to the depositional strike of the coastline and with an average of 7 cores per transect (Figure 1A). The transects extended lakeward from the Calumet Beach to the landward side of the Toleston Beach. This core distribution was found to adequately sample the basin. An additional 7 cores were collected along a transect parallel to the long axis of the wetland in the western part of the study area (Figure 1A). This transect crosses a 3- to 4-ft-high (0.9- to 1.2-m) peat mound in Cowles Bog.

Cores were returned to the laboratory to be split, described, sampled, and photographed. Because stratification is difficult to recognize in unconsolidated deposits, latex peels were made for many of the cores to enhance sedimentary structures. Genetic facies were defined on composition, texture, sedimentary structures, upper and lower contacts, and associated sedimentary deposits.

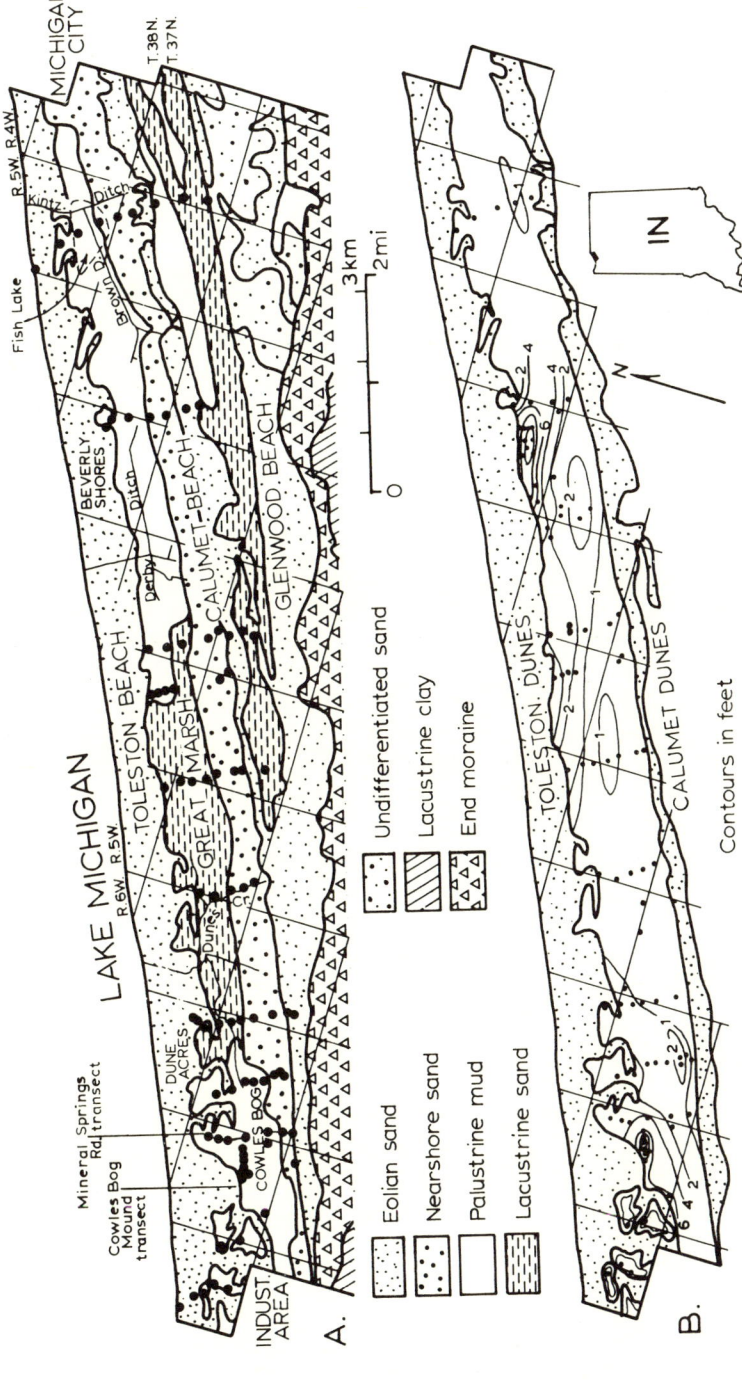

Figure 1. Maps showing the study area. A. Geologic map of the surficial deposits along the southeastern shore of Lake Michigan showing vibracore locations (dots), geomorphic features, and drainages. B. Isopach map of the palustrine deposits in the wetland between the Calumet and Toleston dunes. (Note: 0.3048 ft = 1 m)

RESULTS

Sedimentary Deposits

Surficial and subsurface deposits of the Great Marsh are composed of interbedded palustrine, back-barrier lacustrine, and eolian deposits. Although highly variable in their distribution, sediments of these environments are common throughout the study area. Descriptions of their physical characteristics are given below.

Palustrine

Palustrine deposits consist of fibric, sapric, and humic (organic mud) peat with intercalations and scattered grains of sand. Many horizons within the peat are calcareous, and gastropod and pelecypod shells are common. Throughout the study area, the palustrine deposits are generally less than 2 ft (0.6 m) thick, but at Cowles Bog and south of Beverly Shores, Indiana, the palustrine sequence reaches thicknesses of 10 ft (3.0 m) or more (Figure 1B).

Most cores taken in the Great Marsh are from Cowles Bog (Figure 1A). Here, there is an upward change in the palustrine deposits from humus peat to sapric peat and finally fibric peat (Wilcox et al., 1987). Accompanying the peat sequence is a progression in floral types from an assemblage of lily and water shield to an assemblage of tamarack, jack pine, speckled alder, holly, fern, and sphagnum moss (Futyma, 1985, and this volume). Miller and Thompson (1986) also reported an upward faunal succession in the molluscs at Cowles Bog. A pulmonate gastropod assemblage dominates the humic and sapric peat at the base of the sequence. Terrestrial gastropods are increasingly abundant in the fibric peats of the upper half of the section. Thin interbeds of sapric peat and their associated fauna, however, do occur in the lower part of the fibric peat section.

Back-barrier lacustrine

Back-barrier lacustrine deposits primarily consist of 0.25- to 2.5-ft-thick (0.08- to 0.8-m) beds of calcareous clay and marl with interbeds of sand. The calcareous clay is organic-rich and highly fossiliferous, and it contains disseminated and compacted plant debris and a fauna of gastropods, pelecypods, ostracods, turtles, and fish (Gutschick and Gonsiewski, 1976). Calcareous clay is exposed along the Lake Michigan shoreline near Michigan City, Indiana, for about 2 mi (3.2 km)(Winkler, 1962). Similar deposits also were collected at the bottom of cores taken in the extreme western and eastern parts of the study area.

The marl is composed of light-gray, fibrous and nonfibrous micrite, silty micrite, and sandy micrite. The mud has a well-developed horizontal parallel-lamination and contains scattered prosobranch gastropods and pisidid clams (Miller and Thompson, 1986). Stratification is defined by light- and dark-colored laminae (organic content?) and by intercalations of fine-grained sand. Marl beds are laterally discontinuous and can be traced between cores in only a few places.

Interbedded with the calcareous clay and marl are horizontally-laminated and ripple cross-laminated beds (0.5 to 3.0 ft thick, 0.15 to 0.7 m) of fine-grained sand. Laminae are defined by alternations in grain size but also by intercalations of fibric and sapric peat. In many places the peat may be laterally projected into marl, implying that the marl beds are surrounded by peat.

Eolian

Eolian sediments are composed of well-sorted fine-grained sand with a low- to high-angle parallel cross-lamination. Beds are 1 to 15 ft (0.3 to 4.6 m) thick, and laminae within the beds are defined by heavy minerals or small changes in grain size. Eolian deposits associated with the Great Marsh are confined to the parabolic dunes on the landward side of the Toleston complex.

Internal Architecture of the Great Marsh

The Calumet Beach extends northward beneath the Great Marsh, and this dune and beach complex and older glacial deposits (Wedron Formation) are the foundation on which the wetland sediments accumulated (Figure 2). The Calumet complex also forms the southern margin of the Great Marsh basin that is roughly defined by the dune-beach contact as illustrated in Figure 1A. The northern margin of the Great Marsh basin is less easy to define because sediments of the Great Marsh are interbedded with and also overlain by sediments of the Toleston complex. The northern margin of the basin is arbitrarily selected as the southern extent of the Toleston dune deposits. The wetland basin is asymmetric throughout the study area. Its deepest part is along the northern margin of the basin and gradually shallows over the Calumet complex to the south (Figure 2).

An idealized vertical sequence of Great Marsh sediments consists, from bottom to top, of calcareous clay and lacustrine sand, marl and lacustrine sand, marl and humic peat, sapric peat, and fibric peat. This sequence is not everywhere complete in the basin because changes due to the interbedding of Great Marsh sediments and eolian deposits and onlap of the Calumet complex along the margins of the basin. Interbedding occurs primarily in the northern part of the wetland where Toleston dune deposits are intercalated with lacustrine deposits and compose the major part of the preserved sequence. In the southern part of the basin where Great Marsh sediments overlie the Calumet complex, lacustrine sediments are absent, and the palustrine deposits grade southward into organic-rich soils.

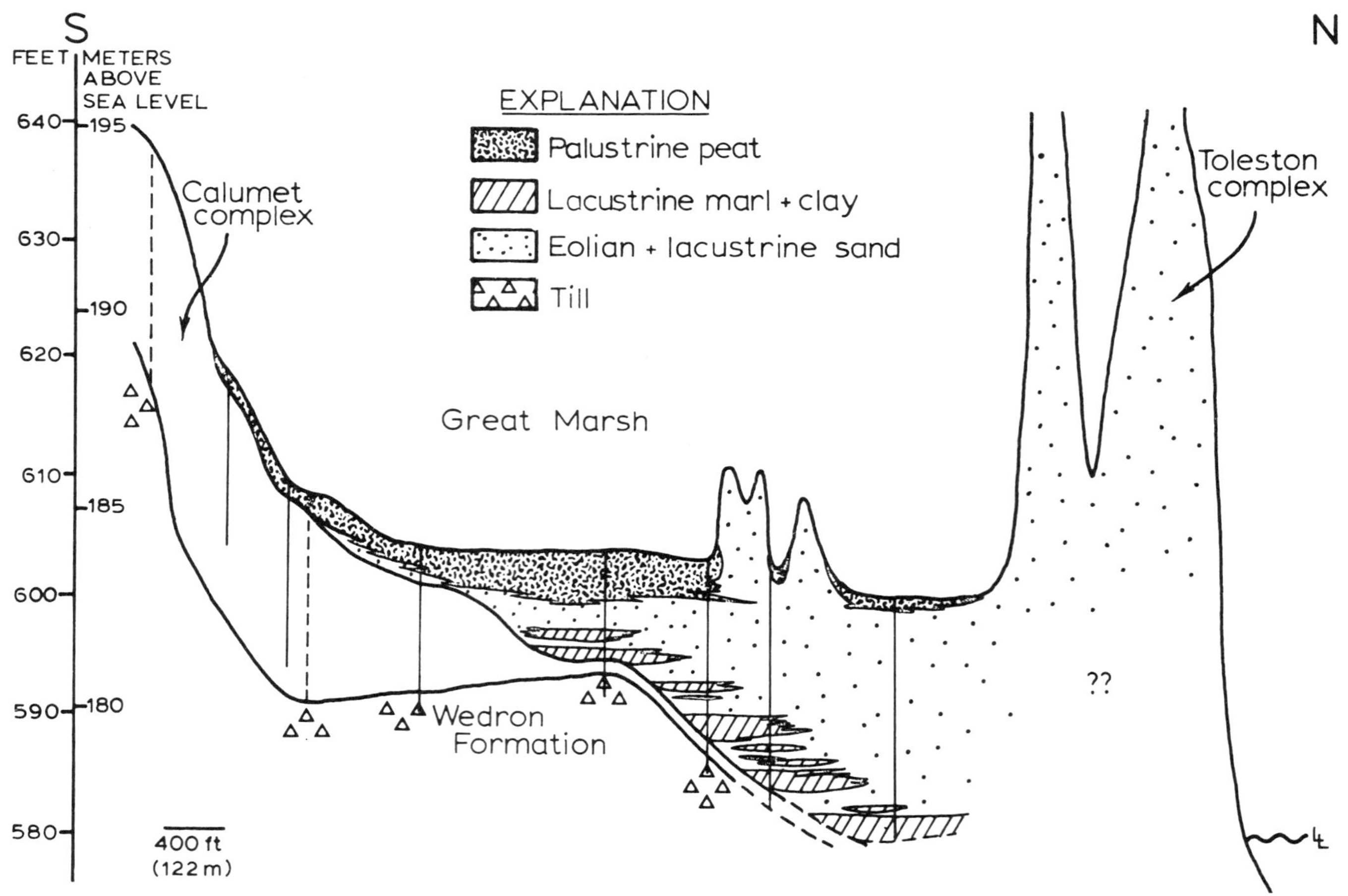

Figure 2. N-S cross section along Mineral Springs Road. (See Fig. 1A for location of transect.) Solid vertical lines are vibracores, and dashed vertical lines are water wells. Triangles at the base of a core or well indicate that it has penetrated till of the Wedron Formation.

A complete sequence of wetland sediments occurs at Cowles Bog. Here, calcareous clays, in the deepest part of the basin, become progressively less clay rich upward in the section and to the south, where they overlie the Calumet complex (Figure 2). In the northern part of the basin, the lacustrine clay, sand, and marl are overlapped by Toleston dune deposits that migrated east and east-southeast over the lacustrine sediments. The entire sequence is capped by 2 to 4 ft (0.6 to 1.2 m) of palustrine deposits.

The depositional-strike section across the peat mound in Cowles Bog (Figure 3) shows a similar sequence, although the marl and fibric peat are considerably thicker beneath the mound. The lateral discontinuity of the marl and the variable thickness of the palustrine deposits are clearly illustrated in this section.

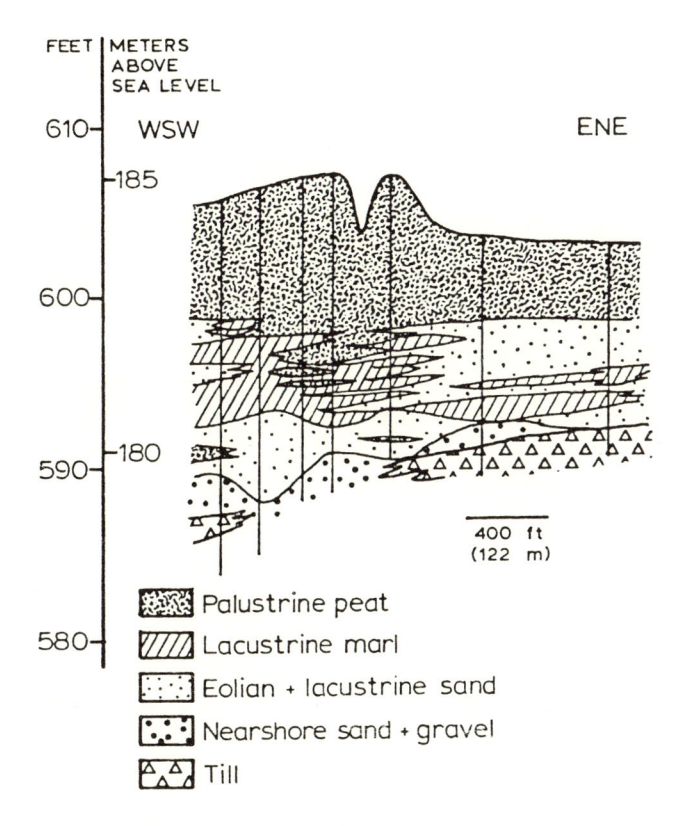

Figure 3.- WSW-ENE cross section across a peat mound in Cowles Bog. (See Fig. 1A for location of transect.) Solid vertical lines are vibracores.

Variation in the distribution of sediment types also occurs along the long axis of the Great Marsh. The peat and lacustrine sediments are the thickest at Cowles Bog and south of Beverly Shores (Figure 1B). They thin toward the Indiana Dunes State Park and are thinnest along Dunes Creek where the peat is interbedded with fluvial deposits.

Few cores were collected east of Beverly Shores. The cores taken from the wetland consist mostly of lacustrine sand overlain by 2 ft (0.6 m) of palustrine deposits. These sediments occupy the basin of Fish Lake, a presettlement lake just west of Michigan City that was drained in the late 1800's (Cook and Jackson, 1978). Some calcareous clay, however, occurs north of the Fish Lake deposits where beds of clay are interbedded with eolian deposits in exposures along the shoreline (Winkler, 1962; Gutschick and Gonsiewski, 1976; Thompson, 1986).

DISCUSSION: CONTROLS ON WETLAND SEDIMENTATION

Major factors that control the type and intensity of the available processes in a wetland are: the origin of the basin and adjacent land area, the position of the basin in the coastal complex, predepositional and intrabasinal topography, drainage, water chemistry, hydraulic connection with the open coast, surface water, and ground water, floral and faunal content, climate, and age (Badyakov, 1986; Weller, 1978). Careful consideration of the influence of these factors through time is important in understanding the evolution of the basin. Excluding time, few of these factors are independent of the others, and knowledge of their significance cuts across scientific disciplines. Consequently, diverse backgrounds are needed to fully understand the history and maturation state of the wetland.

In this section, I will discuss briefly the origin of the Great Marsh and the influence of the open lake, ground water, predepositional topography, and drainage as they pertain to the development of the wetland. Details of the chemistry, hydrology, and floral content of the Great Marsh are covered by Wilcox et al. (1984), Wilcox et al. (1987), and elsewhere in this volume. (See Wilcox; Shedlock et al.; and Futyma.)

Origin

The Calumet Beach began to form about 11,800 years ago (Hansel et al., 1985) and prograded lakeward for a little more than 1,500 years. After this time, deglaciation of the northern outlets to the Lake Michigan basin lowered lake level 180 ft (60 m) below the present altitude of the lake (Bluckley, 1974). This low stage lasted for about 4,500 years (Hansel et al., 1985) and was concurrent with the Hypsithermal, a climatic interval of warmer and drier conditions that affected the Midwest during early Holocene time (Bartlein and Webb, 1982).

Isostatic rebound of the northern outlets and cooler and wetter conditions following the Hypsithermal produced a rise in lake level. Along the southern shore of Lake Michigan, a beach ridge developed offshore of the present position of the Indiana Dunes, and this ridge migrated landward as lake level rose to the altitude of glacial Lake Nipissing (Thompson, 1986). This beach ridge formed the northern margin for a wetland that was established on the south side of the barrier about 7,800 years ago. The pattern of sedimentation within the back-barrier basin is typical of north-temperate-zone lakes (see Fouch and Dean,

1982) but was strongly influenced by a continued rise in lake level in the Lake Michigan basin.

Influence of the Open Lake

The first sediments that were deposited in the wetland were calcareous clay and lacustrine sand. They accumulated in lakes that ponded landward of the beach ridge. The lake that formed just west of Michigan City was quite large, extending more than 2 mi (3.2 km) along the shoreline (width is unknown due to present erosion), and may have persisted for about 1,500 years. Clay and sand were supplied to the back-barrier lakes by overwash and eolian transport (Thompson, 1986). As the the beach ridge vertically aggraded and as dunes developed on the beach ridge, the amount of overwash into the back-barrier basin, and therefore the supply of detrital clays to the lakes, diminished. Consequently, the lacustrine deposits became less clay rich and more calcareous as the basin was filled. These marl ponds lacked the lateral continuity of the earlier lakes and may have been maintained by local supplies of ground water.

Lake-level rise in the Lake Michigan basin continued throughout this early history of the wetland and caused a slight landward translation of the beach ridge and encroachment of parabolic dunes into the wetland. The marl ponds migrated landward and vertically over the lakeward edge of the Calumet complex as the dunes extended into the basin. As lake level stabilized at 603 ft (184 m) above mean sea level and as vegetation spread from the margins of the basin and across the entire wetland, the marl ponds were filled. Except for Fish Lake, a peatland was formed throughout the study area. The base of the peat sequence dates at 3,600 years B.P. (Wilcox et al., 1987).

Drainage Influence

Dunes Creek alone drained the wetland during its early history. The inception of Dunes Creek is unknown, but this creek has influenced the development of peat in the Indiana Dunes State Park. The peat in the state park is thinner and more decomposed than elsewhere in the basin, presumably as a result of the moderate drainage that Dunes Creek produces. Apparently, drier conditions near Dunes Creek did not permit the development of large quantities of peat.

Topographic Influence

The morphology of the Calumet Beach is a major control on paludal sedimentation. Peat deposits thin with onlap of the Calumet nearshore and thicken considerably north of the complex. This thickening northward is most prevalent in the eastern part of the Great Marsh (Figure 1B), where thicknesses of 10 ft (3.0 m) or more are attained south of the dunes of the Toleston complex. Relief on the surface of the Calumet complex also causes local areas of peat thinning and thickening. Several areas of peat thickening are illustrated along the southern margin of the Great Marsh in Figure 1B.

Ground Water Influence

Ground water is a principal water source of the Great Marsh. Many of the flow paths are local phenomena and stem from the dune and beach complexes and the lake. These small-scale flow systems are not simple, and local variations due to inhomogeneities in beds and rapid changes in stratigraphy are common (Shedlock et al., in review).

Larger scale flow paths originate in the moraines south of the study area (Shedlock et al., this volume). These deep-seated systems are not influenced by yearly changes of rainfall in the basin and continuously supply water to the wetland. The deep aquifers can have local influence. At Cowles Bog, for example, a breach occurs in the till layer below the peat mound (Figure 3). This breach permits the upward discharge of water into the wetland that is saturated with respect to calcite (Shedlock et al., in review). The thick marl and peat under the mound are a product of this upward discharge of water. They are the result of the high calcium carbonate content of the fluid and continued supply of water to the wetland.

CONCLUSIONS

Factors that control sedimentation in coastal wetlands are highly variable and can change greatly from wetland to wetland and even during the development of a single wetland. Understanding the significance of these factors requires a broad background into many scientific diciplines. The most basic of these disciplines are sedimentology and stratigraphy. They provide a framework for the findings of other fields.

Vibracores were used in the Indiana Dunes to define the internal architecture of the Great Marsh. The subsurface sediments and their distribution in the Great Marsh are similar to the pattern of sedimentation that occurs in most north-temperate-zone lakes. Variations in this model are due to a rise in lake level in the Lake Michigan basin, local perturbations in ground-water flow paths, predepositional topography on which the wetland developed, and distribution of drainage systems in the wetland basin.

LITERATURE CITED

Badyakov, D.D., 1986. Ancient shorelines as indicators of sea level. Journal of Coastal Research 2:147-157.
Bartlein, P.J. and T. Webb III, 1982. Holocene climatic changes estimated from pollen data from the northern midwest. In: Quaternary History of the Driftless Area, J.C. Knox (Editor). Wisconsin Geological and Natural History Field Trip 5, pp. 67-82.

Bluckley, S.B., 1974. Study of post-pleistocene ostracod distribution in the soft sediments of southern Lake Michigan. Ph.D. Thesis, University of Illinois, Urbana, Illinois, 189p.

Cook, S.G. and R.S. Jackson, 1978. The Bailly Area of Porter County, Indiana: The Final Report of a Geohistorical Study Undertaken on the Behalf of the Indiana Dunes National Lakeshore. Robert Jackson and Associates, Evanston, Indiana, 110p.

Finklestein, K. and D. Prins, 1981. An inexpensive portable vibracoring system for shallow-water and land application. Coastal Engineering Research Center, C.E.T.A. 81-8, 15p.

Fouch, T.D. and W.E. Dean, 1982. Lacustine environments. In: Sandstone Depositional Environments, P. A. Scholle and D. Spearing (Editors). American Association of Petroleum Geologists Memoir 31, pp. 87-114.

Frey, R.W. and P.B. Basan, 1978. Coastal salt marshes. In: Coastal Sedimentary Environments, R. A. Davis, Jr. (Editor). Springer-Verlag, New York, pp. 101-169.

Futyma, R. P., 1985. Paleobotanical studies at Indiana Dunes National Lakeshore. National Park Service, Porter, Indiana, USA. 242 pp.

Futyma, R.P., 1988. Fossil pollen and charcoal analyses in wetland development studies at Indiana Dunes National Lakeshore. (this volume)

Gutschick, R.C. and J. Gonsiewski, 1976. Coastal geology of Mt. Baldy, Indiana Dunes National Lakeshore, south end of Lake Michigan. North-Central Section of the Geological Society of America Guidebook for Fieldtrips. Western Michigan University, Kalamazoo, Michigan, pp. 40-90.

Hansel, A.K., D.M. Mickelson, A.F. Schneider and C.E. Larson, 1985. Late Wisconsinan and Holocene history of the Lake Michigan basin. In: Quaternary Evolution of the Great Lakes, P.F. Karrow and P.E. Calkin (Editors). Geological Association of Canada Special Paper 30, pp. 39-53.

Miller, B.B. and T.A. Thompson, 1986. Molluscan succession in cores from the Cowles Bog area, Indiana Dunes, Indiana (abst.). Program and Abstracts for the Ninth Biennial Meeting of the American Quatenary Association, p. 151.

Shedlock, R.J., N. J. Loiacono, and T. E. Imbrigiotta, 1988. Effects of ground water on the hydrochemistry of wetlands at Indiana Dunes, Northwest Indiana. (this volume)

Thompson, T.A., 1986. Post Lake Chippewa transgression deposits in the Indiana Dunes National Lakeshore. In: Quaternary Records of Northeastern Illinois and Northwestern Indiana, A. K. Hansel and W. H. Johnson (Compilers). Field Guidebook for the Ninth Biennial Meeting of the American Quatenary Association, Trip 5, pp. 39-44.

Weller, M.W., 1978. Wetland habitats. In: Wetland Functions and Values: The State of Our Understanding, P. E. Greeson, J. R. Clark, and J. E. Clark (Editors). American Water Resources Association, Minneapolis, Minnesota, pp. 21-234.

Wilcox, D. A., 1988. The necessity of interdisciplinary research in wetland ecology: the Cowles Bog example. (this volume)

Wilcox, D. A., S. I. Apfelbaum, and R. D. Hiebert, 1984. Cattail invasion of sedge meadows following hydrologic disturbance in the Cowles Bog Wetland Complex, Indiana Dunes National Lakeshore. Wetlands 4: 115-128.

Wilcox, D. A., R. J. Shedlock, and W. H. Hendrickson, 1986. Hydrology, water chemistry, and ecological relations in the raised mound of Cowles Bog. Journal of Ecology 74: 1103-1117.

Winkler, E.M., 1962. Radiocarbon ages of postglacial lake clays near Michigan City, Indiana. Science 137:528-529.

EFFECTS OF GROUND WATER ON THE HYDROCHEMISTRY OF WETLANDS
AT INDIANA DUNES, NORTHWEST INDIANA

Robert J. Shedlock[1], Nancy J. Loiacono[2],
and Thomas E. Imbrigiotta[3]

ABSTRACT:

Indiana Dunes National Lakeshore, along southern Lake Michigan, contains wetlands within and between Pleistocene and Holocene dune-beach complexes. Analysis of hydrogeological and hydrochemical data shows that the water chemistry and water balance of the wetlands are influenced by ground-water flow paths of variable length. Spatial changes in hydrochemistry correspond to these different flow paths. For example, flow paths are short in the dune-beach complexes where shallow ground and wetland waters are calcium bicarbonate-sulfate types with dissolved solids and alkalinity concentrations usually less than 300 mg/L (milligrams per liter) and 150 mg/L (as calcium carbonate), respectively. Conversely, ground waters from long flow paths in buried sand aquifers leak upward into a large interdunal wetland (Great Marsh) through the underlying glacial-lacustrine clay. These waters are bicarbonate types with varying proportions of calcium, magnesium, and sodium, and with greater concentrations of dissolved solids (400 to 600 mg/L) and alkalinity (300 to 500 mg/L) than the waters in the dunes. The buried aquifers are recharged below glacial moraines south of the dune-beach complexes. The areal extent and hydraulic connection of these aquifers to the morainal uplands are significant controls on water chemistry and the water balance in the Great Marsh and illustrate the importance of the hydrogeologic setting of wetlands.

INTRODUCTION

The southern shore of Lake Michigan contains a variety of wetlands within and between several major dune-beach complexes (Figure 1) that were deposited inland when levels of ancestral Lake Michigan were higher than at present. Much of the modern shoreline between the northwestern cities of Gary and Michigan City is parkland within Indiana Dunes National Lakeshore and Indiana Dunes State Park. This reach of shoreline (hereafter referred to as Indiana Dunes) also contains several small residential communities and several major industries, including three steel mills, a coal-fired electric power plant, and the Port of Indiana--a major Great Lakes port.

---

[1] U. S. Geological Survey, 208 Carroll Building, 8600 LaSalle Road, Towson, Maryland 21204
[2] U. S. Geological Survey, 5957 Lakeside Blvd., Indianapolis, Indiana 46278
[3] U. S. Geological Survey, Mountain View Office Park, 810 Bear Tavern Road, Suite 206, West Trenton, New Jersey 08628

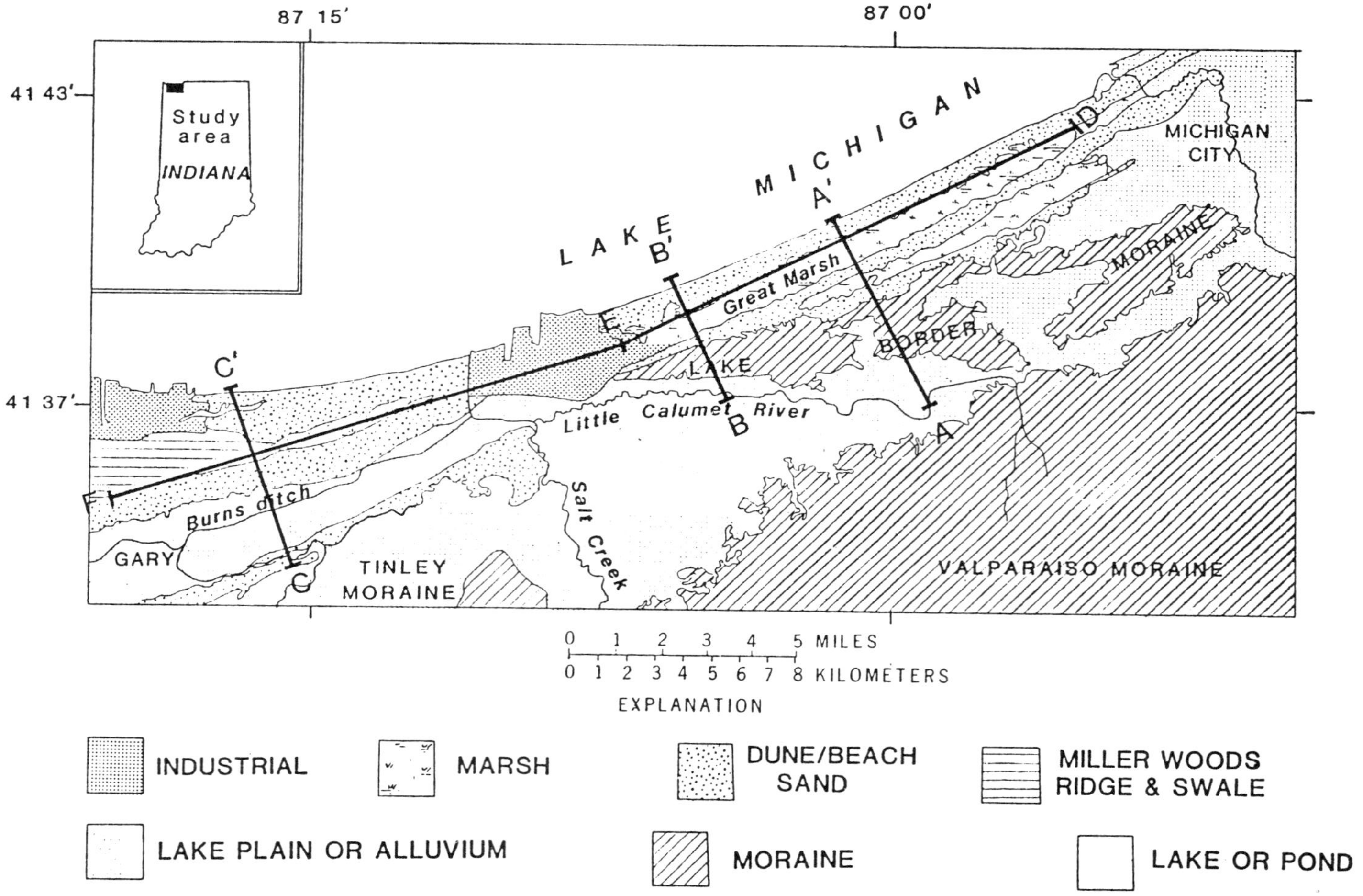

Figure 1. Surficial geology of Indiana Dunes and surrounding area and location of geologic cross sections.

The management and preservation of the dune-wetland ecosystem in such an urban and industrial setting requires an understanding of the physical conditions and processes that control the hydrology and water quality of the area. For this reason, the U. S. Geological Survey and the National Park Service have been conducting cooperative studies of the hydrology, geology, and water quality of the lakeshore area since 1973. One of these studies has been a broad investigation of the hydrogeologic setting of the lakeshore. The purpose of this study was to define the geometry, flow regime, and water quality of the glacial-drift aquifer system underlying the dunes and wetlands (Shedlock, 1983). Other studies have focused on ground-water chemistry and/or flow in smaller areas of this region, including the eastern end of the national lakeshore (Shedlock and Harkness, 1984), the Cowles Bog area (Cohen and Shedlock, 1986; Wilcox et al., 1986), and the central part of the Great Marsh--the largest interdunal wetland in the region (Loiacono, 1986).

The purpose of this paper is to illustrate relations between hydrochemical variations in the wetlands and ground-water flow paths in the glacial drift and surficial sand aquifer system. The interpretations in this paper are based on sedimentological, geophysical, and hydrochemical data from a network of test holes and wells installed between 1979 and 1986. Ground-water flow paths of regional, intermediate, and local scale, in the terminology of Toth (1963), and the water chemistry associated with these flow paths will be described. This discussion is intended to demonstrate that the chemistry of shallow ground water in and around wetlands and, in some cases, the wetland surface water depends on the position of that particular area in the ground-water flow regime. The intervening swales are either ponds or wetlands, depending on their depth. Most of the wetlands in the study area are in lowlands in the lacustrine and aeolian sands. Although wetlands also are in depressions in the morainal uplands to the south and in lowlands along streams (riverine wetlands), the discussion here deals only with wetlands in the lacustrine and aeolian sands.

STUDY AREA

Surficial Geology and Physiography

Most of the national lakeshore is within a physiographic province known as the Calumet Lacustrine Plain (Schneider, 1966). The surficial sediments are lacustrine, beach, dune, and paludal deposits formed in the proglacial lake environment between the ice front and the northern margin of the Valparaiso moraine (Figure 1)--the major end-moraine in the region formed during Wisconsin glaciation. The Calumet Lacustrine Plain also contains surficial tills, such as the Lake Border, that are younger than the Valparaiso moraine and which formed during readvances of the glacial ice into the Lake Michigan basin. The history of the proglacial lake that formed the Calumet Lacustrine Plain, known as Lake Chicago, is described by Leverett and Taylor (1915), Bretz

(1951, 1955), and Hough (1958). Alternatives to the lake-level models in these older works are discussed by Hansel et al. (1985).

Several distinct, well-preserved dune-beach complexes were formed in this physiographic province when the proglacial lake levels were higher than the present mean level of Lake Michigan. These complexes are now stabilized by vegetation and extend across the study area roughly parallel to the modern shoreline of Lake Michigan (Figure 1). Most of the lowlands in this province are wetlands whose different geomorphic settings are illustrated in Figure 2. The largest wetland regions are in the lowlands between the major dune complexes (interdunal wetlands). Wetlands of much smaller areal extent are found in lowlands within the dune complexes (intradunal wetlands). In the western part of the study area, a region known as Miller Woods consists of a series of long, low sand ridges that parallel each other.

Aquifer System

The unconsolidated sediments in the study area are glacial, lacustrine, and aeolian deposits of Pleistocene and Holocene age. The sediments range in thickness from 27 to 90 m (meters) and rest on an erosional surface of bedrock that consists of Mississippian and Devonian shale and carbonate rocks. The surficial deposits are dune, beach, and lacustrine sands that are overlain by peats in the wetlands. At some locations, lenses of calcareous clay of lacustrine and paludal origin separate the peat and sand and/or are interbedded with the sand. These surficial deposits are underlain by a complex sequence of glacial and lacustrine sediments that consist mainly of silty clays with lenses and sheets of sand and sand/gravel. Most of the silty clays are probably glacial tills, although some may be of lacustrine origin. These clays function as confining layers for the buried sand bodies that form the major confined aquifers in the lakeshore area. The glacial-lacustrine deposits form a complex, confined, aquifer system in which the spatial distribution of the buried aquifers varies significantly along the lakeshore (Figure 3).

The buried aquifers are most important in the eastern half of the study area. Here the surficial aquifer below the interdunal wetlands is generally thin (less than 6 m of saturated thickness) because the till at the top of the glacial-lacustrine complex forms a buried shelf that extends beneath the Great Marsh at relatively shallow depth (Thompson, 1985). This glacial-lacustrine complex contains two major buried aquifers. One underlies the eastern half of the Great Marsh near the base of the glacial drift (the basal sand aquifer). The second is just below the till that forms the top of the glacial lacustrine complex (subtill aquifer). The subtill aquifer is under the entire outcrop area of the Lake Border moraine (Figure 1), but as shown in Figure 3, the subtill aquifer extends into the Great Marsh only in the western half of the Great Marsh. At the western end of the Great Marsh, the top of the glacial-lacustrine complex dips sharply into the subsurface. Farther west, the surficial aquifer

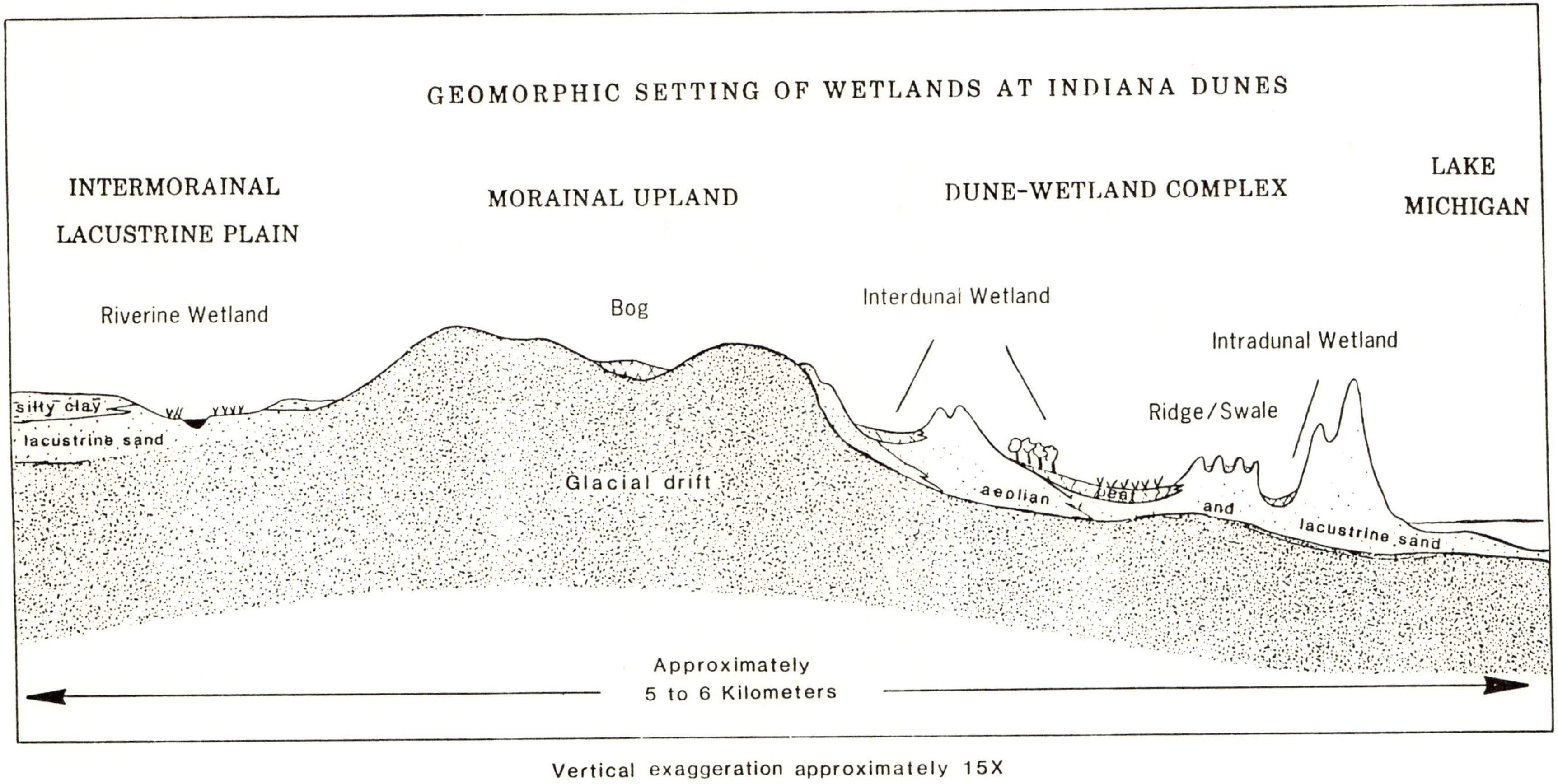

Figure 2. Geomorphic setting of wetlands at Indiana Dunes.

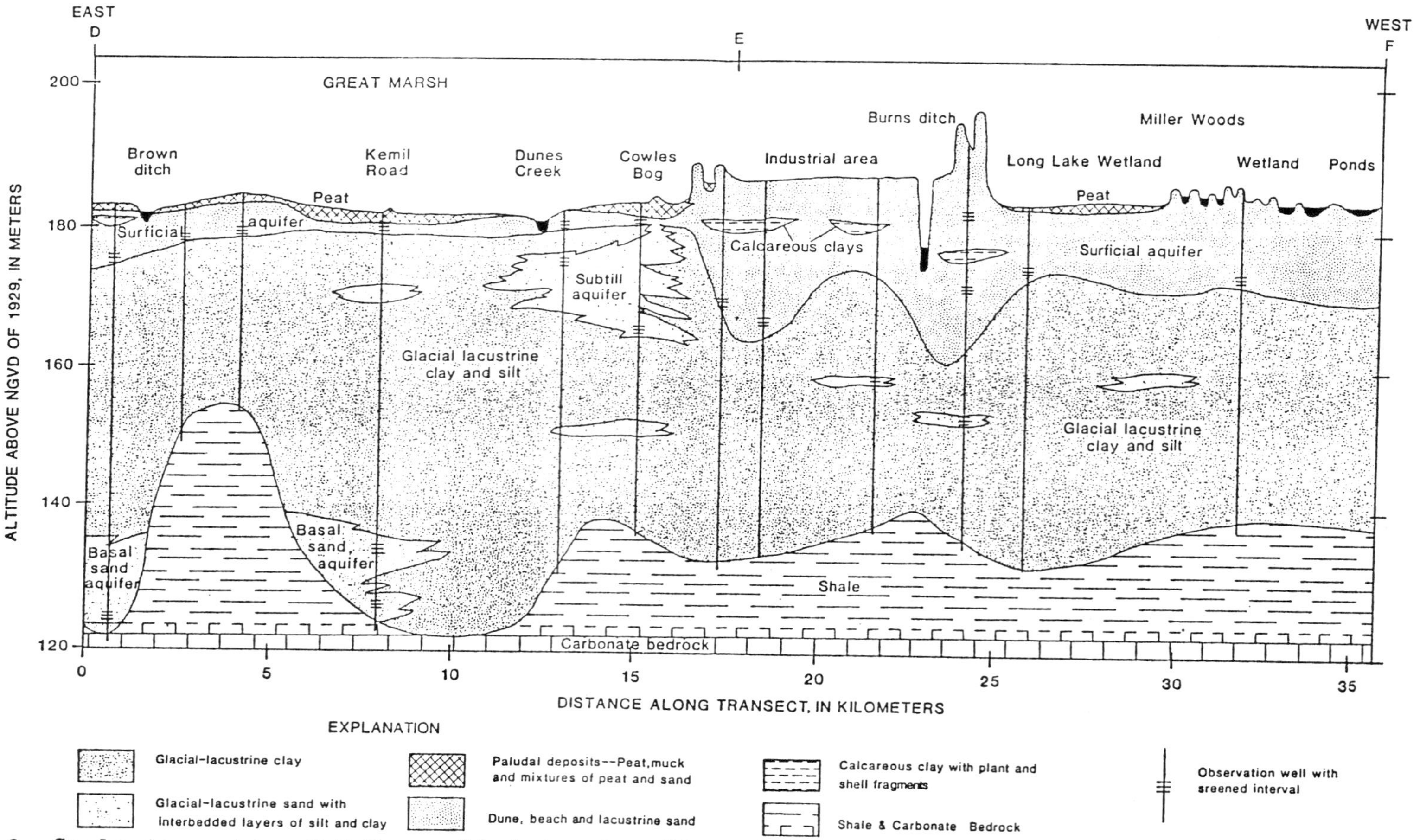

Figure 3. Geologic section D-E-F parallel to shoreline of Lake Michigan showing thickness and distribution of aquifers in unconsolidated sediments. (Trace shown on Figure 1. NGVD of 1929 is derived from a general adjustment of the first-order level nets of both the United States and Canada, formerly called "Mean Sea Level.")

becomes thicker, and, although its lower sections are locally confined by lenses of back-barrier clay (clays deposited in lagoons landward of beach barriers), no significant buried aquifers are in the underlying glacial-lacustrine complex.

METHODS

About 60 test holes were drilled to the bedrock surface in the study area using mud-rotary methods. One to three observation wells were installed at each of these test-hole sites, depending on the vertical distribution of aquifer layers. Gamma-radiation logs were taken in the deepest well at each site and compared to the drillers' logs to interpret vertical lithologic changes. In addition, eight test holes, 15 to 30 m deep, were later drilled with an auger rig to distinguish tills from lacustrine clays at mid-levels in the glacial drift. Sedimentological and stratigraphic variations in the near-surface sediments (0 to 8 m deep) discussed by Thompson (1985) were determined from over 80 cores obtained with a vibracorer, which was modified from the design of Finkelstein and Prins (1981).

Observation wells finished in buried aquifers were constructed of 10-cm (centimeter)-diameter PVC with 1.5-m long slotted PVC screens. Wells in the surficial aquifer at sites accessible to a drill rig also were constructed of PVC pipe, whereas wells in the surficial aquifer within the wetlands were hand-driven, 5-cm, diameter steel pipe with 0.75-m-long slotted steel screens.

Water samples were collected for chemical analysis from wells in both the surficial and confined aquifer systems. Standing water on the peat surface (wetland water) and pore waters of peat also were collected at well sites in the Great Marsh. Alkalinity, pH, temperature, and specific conductance were measured in the field using the methods recommended by Wood (1976). All samples were analyzed at U. S. Geological Survey laboratories for calcium, magnesium, sodium, potassium, sulfate, chloride, and several minor elements according to the methods of Skougstad et al. (1979).

RESULTS AND DISCUSSION

The geohydrologic section in Figure 4 illustrates the geometry of the aquifer system, hydrochemical types, and directions of ground-water flow along a transect (A-A' in Figure 1) perpendicular to the long axis of the Great Marsh. Flow directions, shown by arrows, from ground-water systems of regional, intermediate, and local scale can be identified in this section. The regional system is probably recharged at water-table highs in the Valparaiso moraine (Figure 1) to the south. This regional flow system extends down through the unconsolidated glacial sediments into the underlying shale and carbonate bedrock. Water in the bedrock flows toward Lake Michigan with a strong component of upward flow into the base of the glacial drift under the lakeshore (Figure 4).

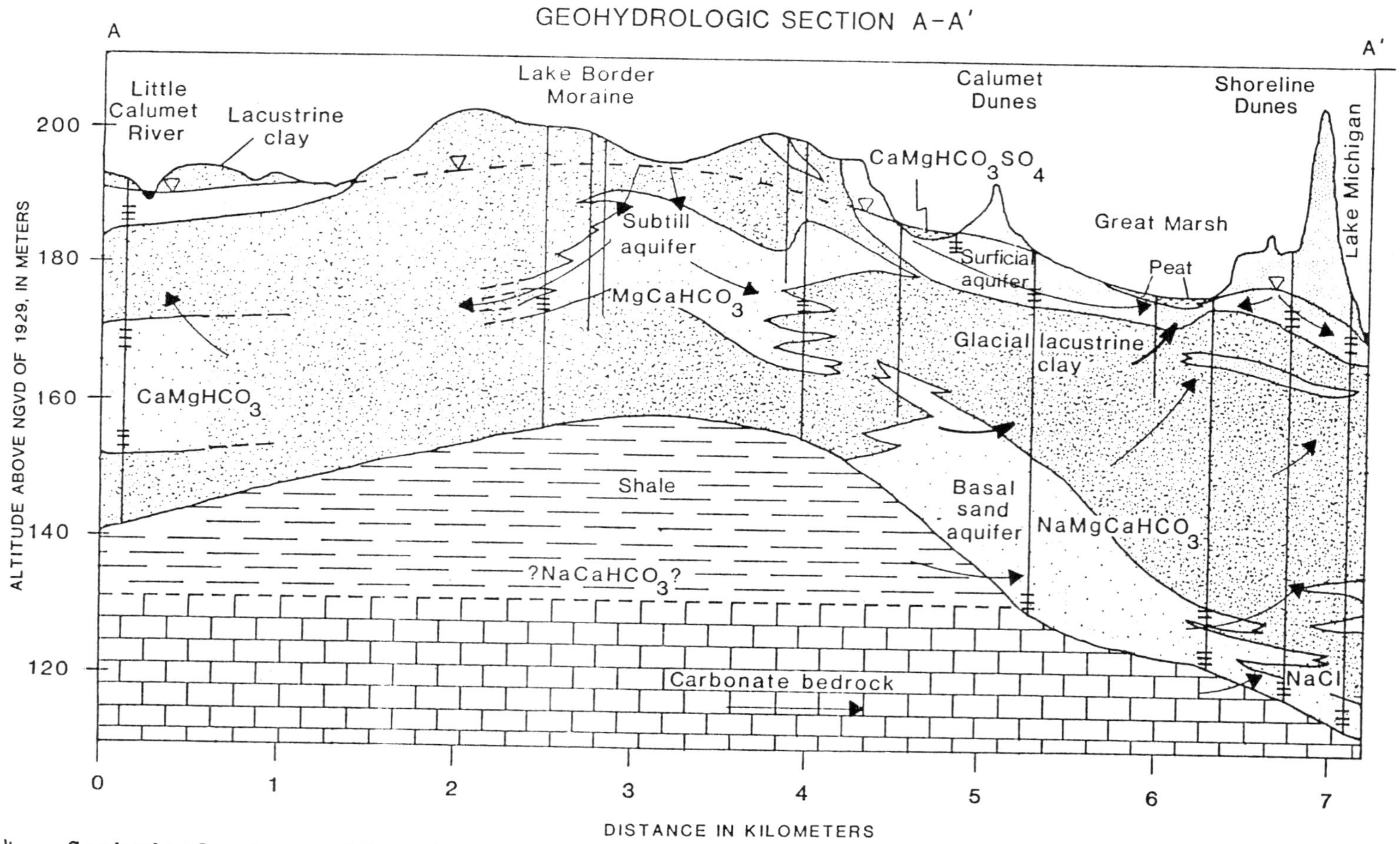

Figure 4. Geohydrologic section in eastern part of Indiana Dunes showing locations and depths of observation wells, thickness and distribution of aquifers, arrows approximating ground-water flow paths, and hydrochemical types of ground water. (Trace shown on Figure 1.)

The arrows that indicate flow from the bedrock into the basal sand aquifer and up through the glacial-lacustrine complex below the Great Marsh represent the distal part of the regional flow system. The upward flow from the bedrock into the glacial drift is inferred from both the vertical distribution of hydraulic head and water chemistry. All wells in the basal aquifer have heads above land surface. In addition, these waters are predominantly hybrid sodium-magnesium-calcium bicarbonate types, with dissolved solids concentrations of 400 to 600 mg/L and alkalinity values of 300 to 500 mg/L (as calcium carbonate). The lone exception is a sodium chloride water (dissolved-solids concentration of 1,200 mg/L) found in the basal sand aquifer along the Lake Michigan shoreline. Because sodium bicarbonate waters are found in the bedrock aquifer west and south of the study area (unpublished U.S. Geological Survey data), the dominance of sodium in waters in the basal sand aquifer is considered evidence of upward flow from the underlying bedrock. The magnesium and calcium in the water of the basal sand aquifer are probably derived either from dissolution of dolomite and calcite in the bedrock or by mixing with downward flowing waters from the subtill aquifer.

The intermediate flow system is recharged in the Lake Border moraine and extends into the underlying subtill aquifer. Water in the subtill aquifer is a magnesium-calcium bicarbonate type in which both dissolved solids concentration and alkalinity range from 300 to 500 mg/L. As shown in Figure 4, water flows both north toward the Great Marsh and south toward the Little Calumet River and discharges to the surficial aquifer by upward leakage through the overlying till. The divide for these divergent flows is probably directly below the water-table high in the Lake Border moraine, which also serves as the shallow ground-water divide.

The local flow systems in the surficial aquifer are recharged in the major dune complexes. Shallow ground water flows from water-table highs in the dunes to the interdunal wetlands where it discharges by seepage to streams and ditches or by evapotranspiration during the growing season. In addition, intensive study of the shallow ground-water system along this transect from the spring to the fall of 1985 (Loiacono, 1986) has shown considerable variability in water chemistry and the position of the water table.

Figure 5 is a detailed cross-section of the shallow ground-water system in the Great Marsh and adjacent dunes along transect A-A' (Figure 1). The water-table profile in Figure 5 shows that the Great Marsh is both a topographic low and a water-table low relative to the adjacent dunes. However, the water table in these dunes does not reflect the topography, and the water-table high in the dunes does not underlie the center of the dune complex. Rather, the water-table high is closer to the Great Marsh on the southern side of the dune complex. In addition, the difference between seasonal high and low water-table altitudes varies across the section. This difference was greatest in the area where the peat is thickest in the Great Marsh. In the dune complex, this

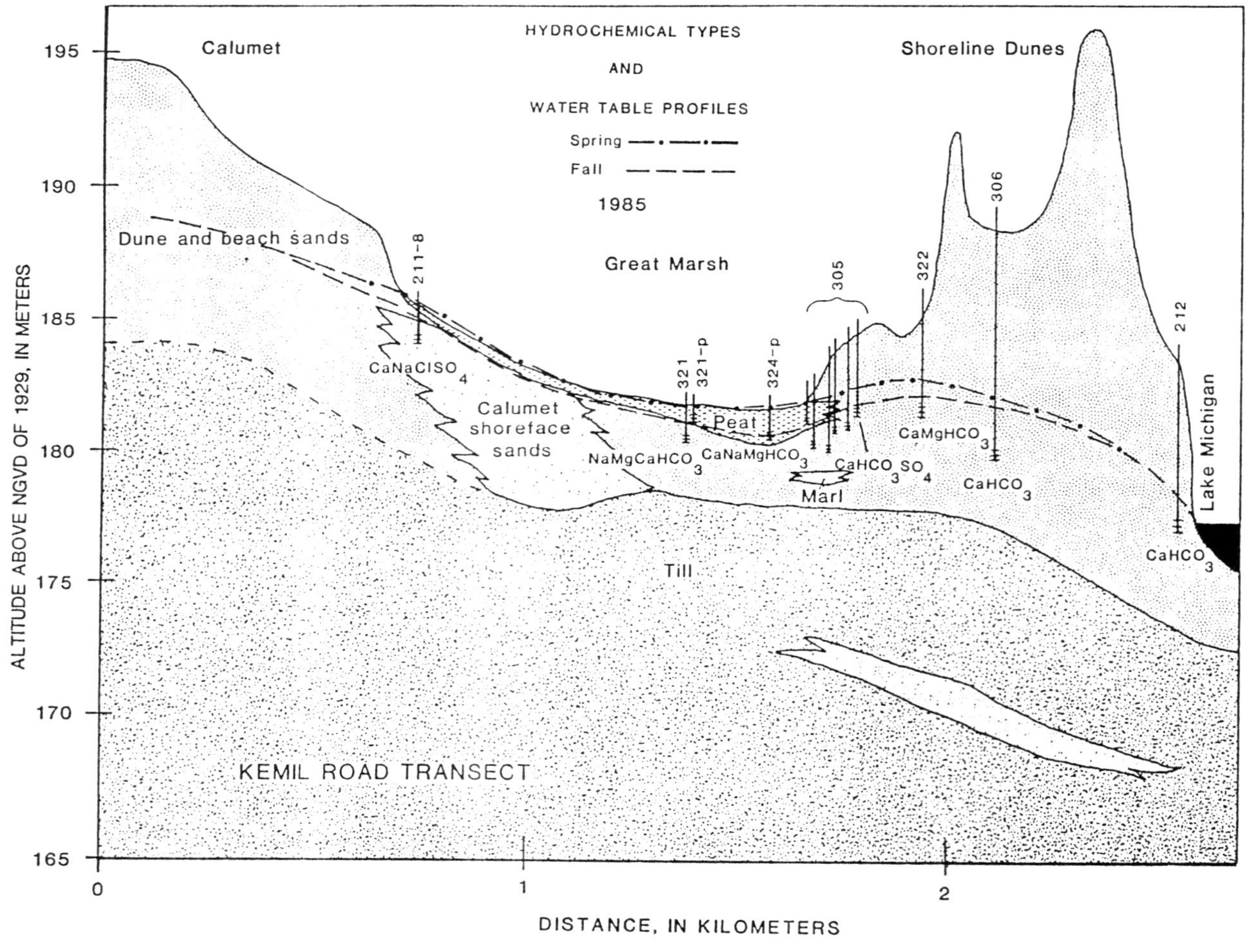

Figure 5. Detailed geologic section of Great Marsh along section A-A' showing observation wells, water-table profile through surficial deposits, and hydrochemical types of ground water. (Trace shown on Figures 1 and 4.)

difference is greater along the margins of the Great Marsh than in the central part of the complexes. The smallest difference is under topographic highs in the dune complexes where the depth to the water table exceeds 6 m. Although the difference was slightly greater in the peat than in the sands of the dune complex at the wetland margin, the amount of recharge to the sand is considered much greater than that to the peat, because the drainable porosity of the peat is much lower than that of the sand. Thus, the differences in the spring and fall water-table profiles suggest that the dune-wetland margin represents the region of the profile most actively recharged by precipitation.

Samples of shallow ground water were collected along this transect to determine if temporal and spatial variations in water chemistry could be related to differences in water-table fluctuations. Several different water types were found, whose relative proportions of major dissolved ions vary considerably along the profile (Figure 5). However, no significant seasonal variations in water chemistry were observed at individual sites other than minor changes in values for total dissolved solids and specific conductance.

The relative proportions of the major dissolved ions for each of the different water types found in the Great Marsh and adjacent dunes and in the underlying basal sand aquifer are graphically represented in Figure 6. Systematic trends are observed in both the anion and cation plots proceeding southward from the shoreline dunes into the Great Marsh. The cations follow a calcium-to-sodium trend from north to south. The anion trend is from bicarbonate to sulfate and bicarbonate and then back to bicarbonate. For example, waters in the topographically-higher sections of the dune complex north of the Great Marsh, where water-table fluctuations were minimal, are mainly calcium bicarbonate types. However, at the dune-wetland margin in the southern part of the shoreline dunes, where greater fluctuations in the water-table were observed, the shallow ground water is a hybrid calcium bicarbonate-sulfate type. The source of the sulfate has not been verified but probably is, at least in part, atmospheric deposition, further indicating that this region is a major recharge area. The waters with this sulfate component tend to have the lowest dissolved solids (less than 300 mg/L) and alkalinity (less than 150 mg/L) concentrations along this profile.

Waters in the peat and in the sands immediately underlying the peat in the Great Marsh are hybrid carbonate types with varying proportions of calcium, sodium, and magnesium. These waters generally have higher dissolved solids and alkalinity concentrations than the shallow ground waters in the adjacent dunes.

Shallow ground water at the southern edge of the Great Marsh is a hybrid calcium-sodium chloride-sulfate type. This water also has lower values of dissolved solids and alkalinity than waters in the center of the Great Marsh. The sodium and chloride components

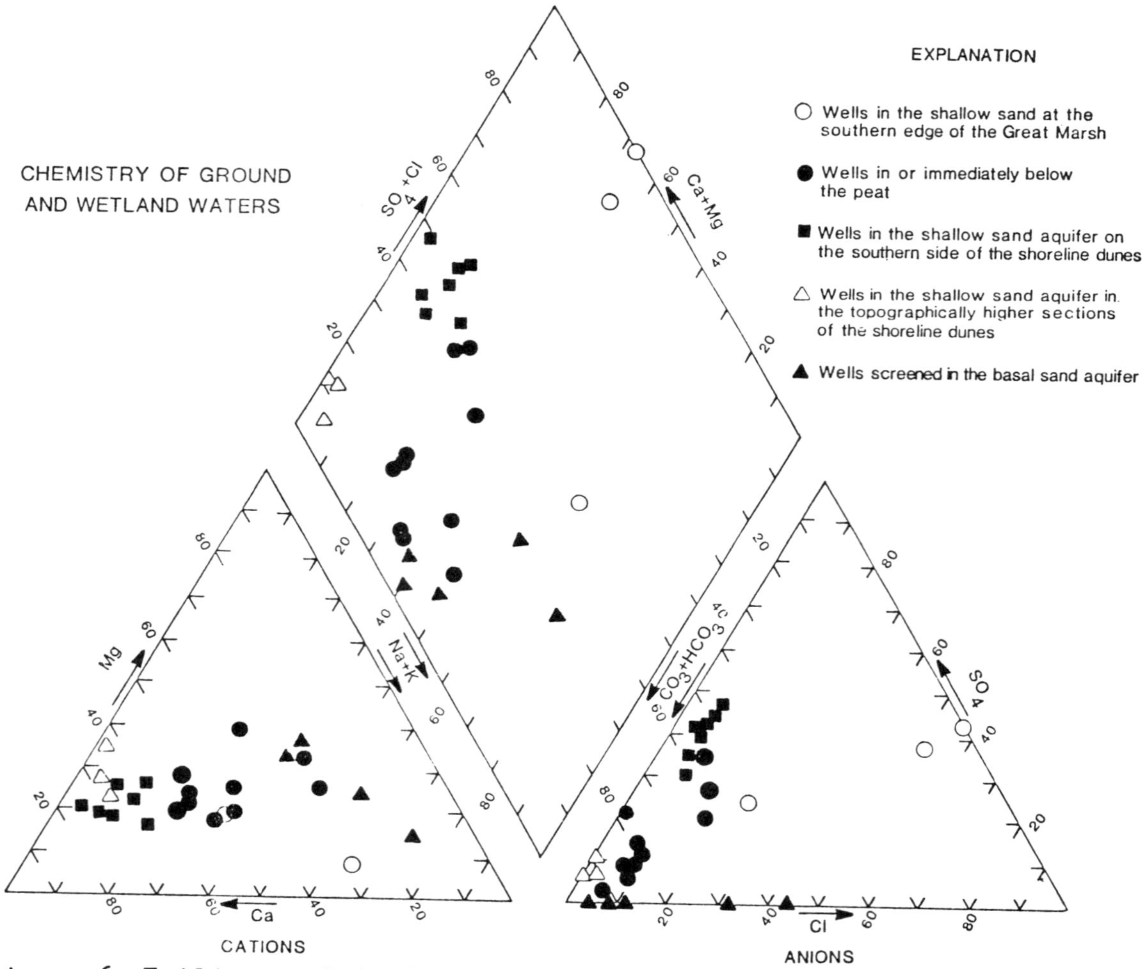

Figure 6. Trilinear plot of chemical composition of waters in Figure 5.

are thought to be from salting of a highway immediately upgradient. Without the sodium chloride in this water, its composition would probably be a calcium bicarbonate-sulfate type, similar to waters in the dune complex at the northern margin of the Great Marsh.

As shown on the plot in Figure 6, the points representing waters in the peat and immediately below it in the surficial sand aquifer are between points representing waters from the shallow ground water in the shoreline dunes at the dune-wetland margin and waters from the basal sand aquifer. This relation is consistent with the flow system in Figure 4, which indicates that the Great Marsh is a zone of discharge for both shallow ground waters from the adjacent dunes and waters from the basal sand aquifer that leak up through the underlying glacial-lacustrine clays.

Upward leakage from an underlying confined aquifer also has been documented farther to the west in the Great Marsh in Cowles Bog (Figure 7), which contains a raised peat mound at the northern edge of the Great Marsh. The hydrology and water chemistry of Cowles Bog are described by Wilcox et al. (1986), whereas palynological changes in the peat in this area are described by Futyma (1985). The pore waters in the peat at the mound are similar in composition to waters in the underlying subtill aquifer, indicating that the peat mound is a fen, in wetland terminology, in spite of the proper name of Cowles Bog. The till layer that serves as the confining layer is breached below the peat mound; the breach allows water from the subtill aquifer to flow up into the overlying sands, marls, and peat.

The waters from the subtill aquifer in this section (Figure 7) are calcium-magnesium and magnesium-calcium types (Figure 8) in which dissolved solids concentrations range from 340 to 420 mg/L and alkalinity from 300 to 370 mg/L. These waters are within the intermediate flow system that is recharged in the Lake Border moraine and discharges either southward to the Little Calumet River or northward to the Great Marsh. However, in the wetlands and shallow ground water in the adjacent dune complexes, the waters are dominantly calcium bicarbonate sulfate types similar in composition to waters within the dune complexes in section A-A' (Figure 5). These waters are derived mainly by recharge from precipitation and are within local flow systems that originate in the dune complexes and discharge either into the Great Marsh or Lake Michigan. The waters within the dune complex in Figure 7 have dissolved solids concentrations less than 300 mg/L, alkalinity less than 150 mg/L, and plot in a different area of the trilinear diagram in Figure 8 than waters from the subtill aquifer and the surficial aquifer in the Great Marsh.

A short distance west of Cowles Bog, the till surface that forms the top of the glacial-lacustrine complex dips abruptly into the subsurface (Figure 3). Farther west of this point, no significant buried aquifers are in the glacial-lacustrine complex. The surficial aquifer is thicker, and local flow systems recharged

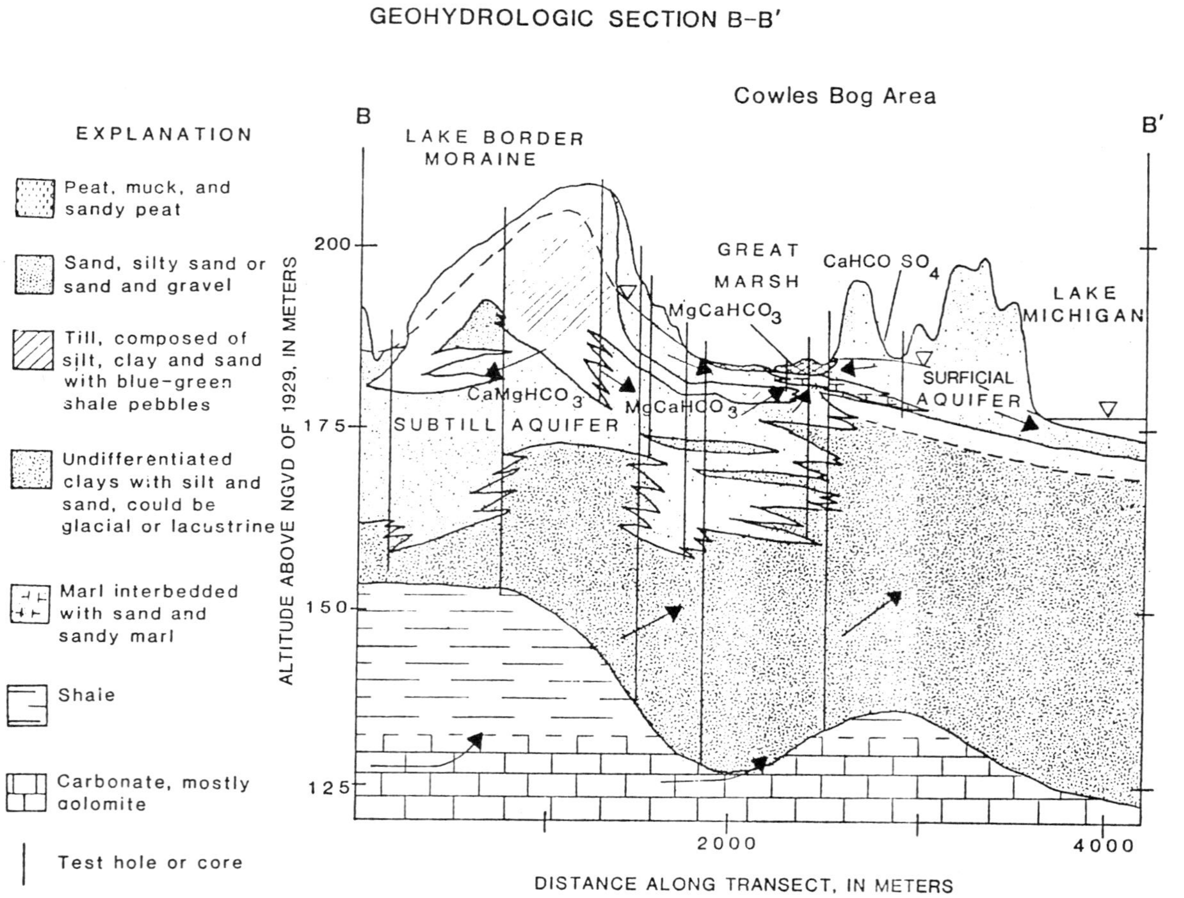

Figure 7. Geohydrologic section in middle part of Indiana Dunes through Cowles Bog area, showing observation wells, thickness and distribution of aquifers, arrows approximating ground-water flow paths, and hydrochemical types of ground water. (Trace shown on Figure 1.)

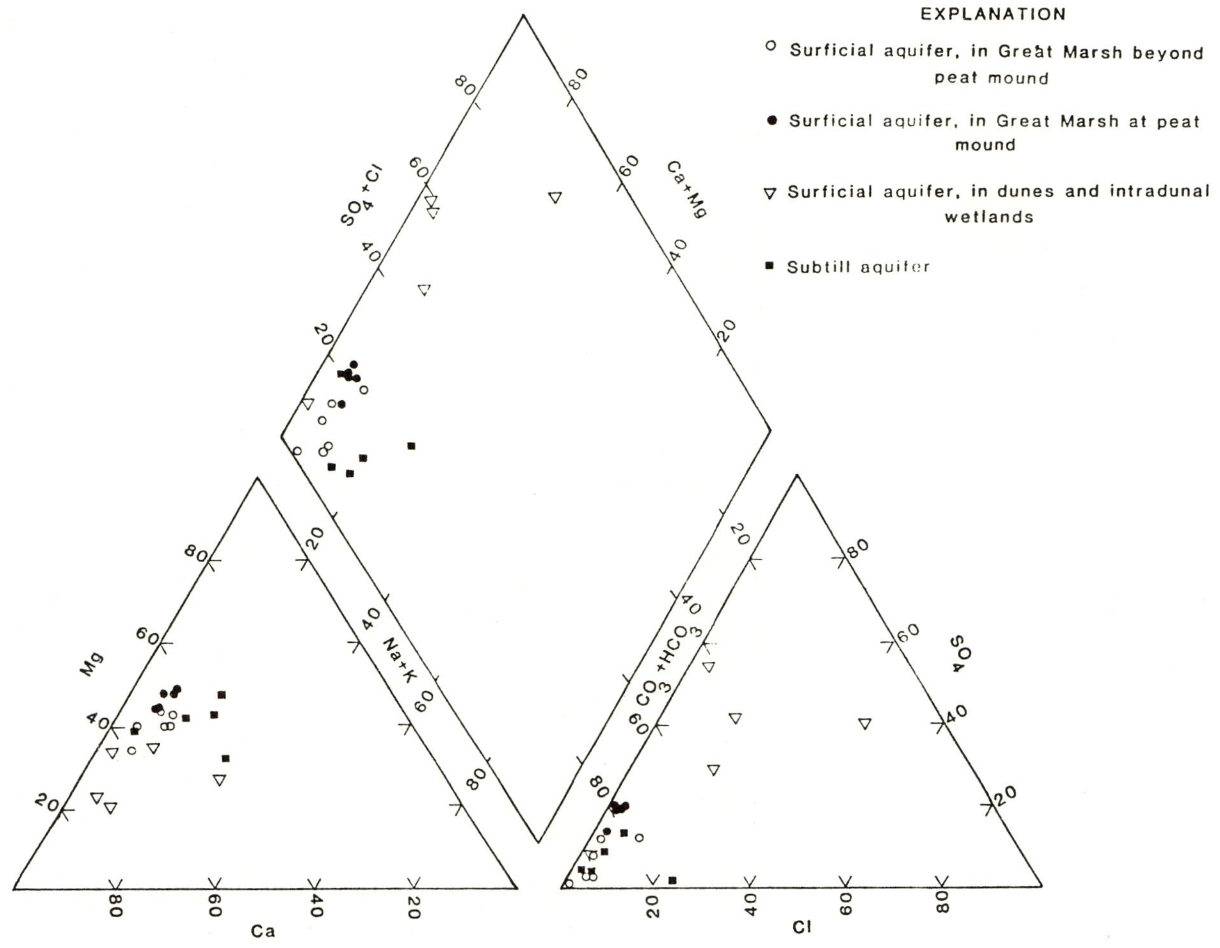

Figure 8. Trilinear plot of chemical composition of ground waters in vicinity of Cowles Bog.

by precipitation are able to penetrate to greater depths. For example, the flow lines for the geohydrologic section in Figure 9 indicate that wetlands and ponds in Miller Woods, at the western end of Indiana Dunes, are fully within the local flow systems. All the shallow ground and wetland waters analyzed from this setting, including several from shallow wells not shown in Figure 9, are calcium sulfate bicarbonate types similar to those found along local flow systems in the dune complexes in the other two sections described earlier. These compositions suggest that none of the wetlands in the western part of the study area are discharge zones for intermediate- or regional-scale flow systems. Brackish sodium chloride and sodium sulfate waters were found in wells screened at the base of the surficial aquifer. These waters may be derived by upward leakage through the glacial-lacustrine clays of water from regional flow paths in the bedrock. However, these brackish waters were not found anywhere near the water table.

CONCLUSIONS

The relation between ground water and wetlands at Indiana Dunes is highly dependent on the position of the wetlands with respect to regional, intermediate, and local ground-water flow systems. The position of a wetland area in the ground-water flow system is influenced greatly by the geometry of underlying aquifer systems in the glacial drift and upper bedrock. The largest wetland here, an interdunal lowland known as the Great Marsh, is underlain by two major buried aquifers in the glacial drift. Ground water from regional and intermediate flow systems, recharged in morainal uplands to the south, discharges into the Great Marsh by upward leakage through the underlying confining clay beds. In addition, ground water from local flow systems, recharged in the adjacent dune complexes, discharges into the Great Marsh.

The upward leakage is driven by the hydraulic connection of the buried aquifers to the morainal uplands. Most of the other wetlands at Indiana Dunes either are not underlain by such buried aquifers or are not in discharge zones of the underlying buried aquifers. Therefore, discharge to these other wetlands is from local ground-water flow systems.

The waters from each of these flow systems have different chemical compositions, and these differences were used along with geologic, geophysical, and hydraulic data to develop conceptual models of the ground-water flow regime. In addition, data from previous sedimentological and paleobotanical studies were used to refine the hydrogeologic framework in areas of local interest such as Cowles Bog. The synergy developed by such an interdisciplinary approach proved invaluable in developing the interpretations in this study. Furthermore, the results of this type of work should prove valuable in evaluating ecological relations in wetlands, many of which are influenced by hydrology and hydrochemistry. A consensus among a growing number of ecologists (D. A. Wilcox,

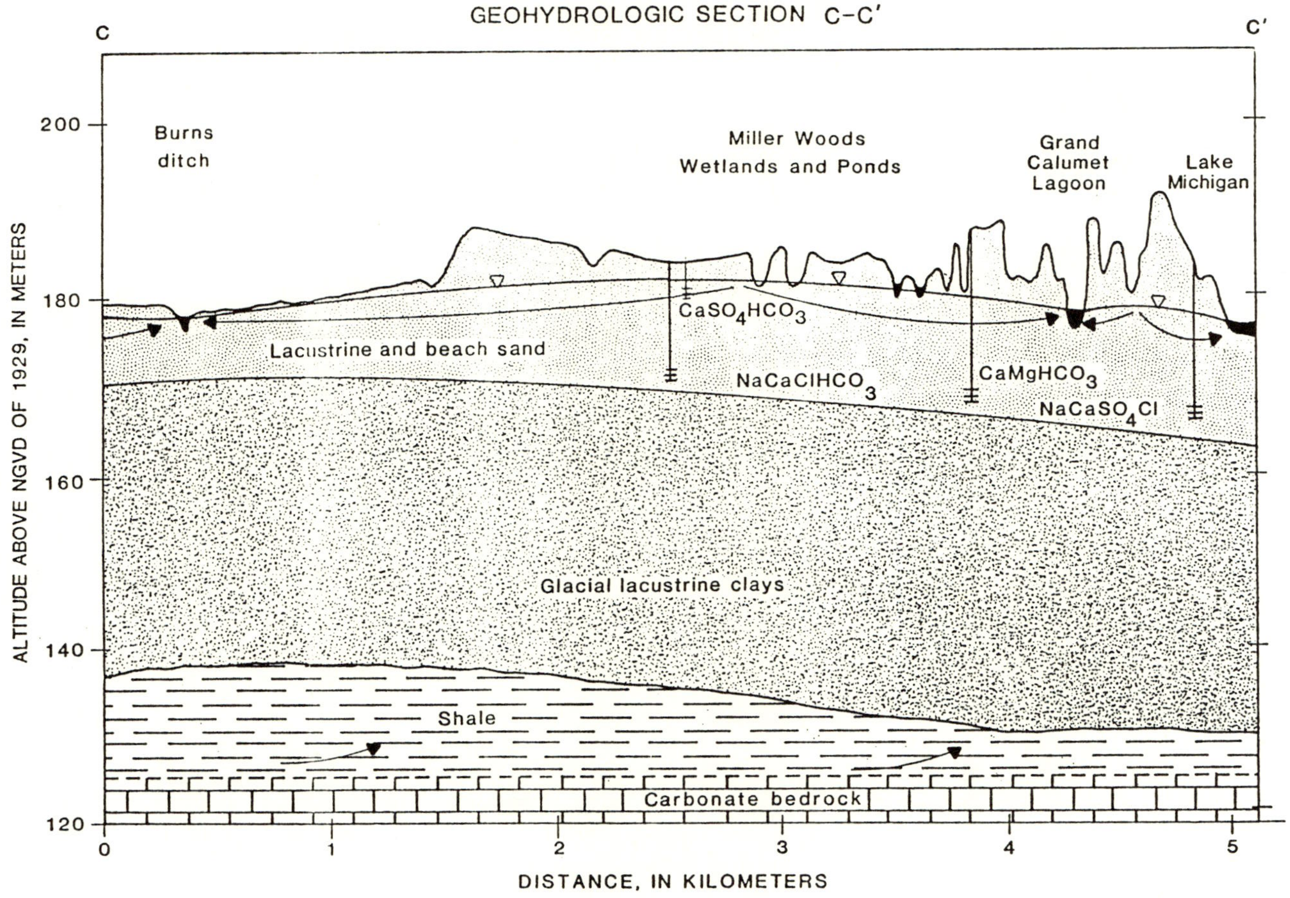

Figure 9. Geohydrologic section of Indiana Dunes showing observation wells, thickness of surficial aquifer and underlying glacial-lacustrine clays, hydrochemical types of ground water, and arrows approximating ground-water flow paths. (Trace shown on Figure 1.)

National Park Service, oral commun., 1986) is that such data are fundamental to understanding ecological processes in wetlands.

ACKNOWLEDGMENTS

This work is the result of cooperative efforts among the U.S. Geological Survey, the National Park Service, and the Indiana Geological Survey. We wish to thank David Cohen, Lee Watson, and Cheryl Silcox, U. S. Geological Survey, for the assistance in both the office and the field. We also wish to thank our colleagues at Indiana Dunes National Lakeshore, particularly Douglas Wilcox and Ronald Hiebert, for their comments and scientific support. Finally, we wish to thank Todd Thompson, Gordon Fraser, and Ned Bleuer of the Indiana Geological Survey for their invaluable help with geologic interpretations.

LITERATURE CITED

Bretz, J. H., 1951. The stages of the Lake Chicago, their causes and correlations. American Journal of Science 249: 401-429.

Bretz, J. H., 1955. Geology of the Chicago region, part II-the Pleistocene. Ill. State Geological Survey Bulletin 65, 132 pp.

Cohen, D. A. and R. J. Shedlock, 1986. Shallow ground-water flow, water levels, and quality of water, 1980-84, Cowles Unit, Indiana Dunes National Lakeshore. U.S. Geological Survey Water-Resources Investigations Report 85-4340, 25 pp.

Finkelstein, K. and D. D. Prins, 1981. An inexpensive portable vibra-coring system for shallow water and land applications. Coastal Engineering Research Center Technical Aid 81-8, 15 pp.

Futyma, R. P., 1985. Paleobotanical studies at Indiana Dunes National Lakeshore. National Park Service, Porter, Indiana, 242 pp.

Hansel, A. K., D. M. Mickelson, A. F. Schneider, and C. E. Larsen, 1985. Late Wisconsinan and Holocene history of the Lake Michigan basin. In: Quaternary evolution of the Great Lakes, Karrow, P. F., and Calkin, P. E., (Editors). Geological Association of Canada Special Paper 30, p. 39-53.

Hough, J. L., 1958. Geology of the Great lakes. University of Illinois Press, Urbana, Illinois.

Leverett, F. and F. B. Taylor, 1915. The Pleistocene of Indiana and Michigan and the history of the Great Lakes. U.S. Geological Survey Monograph 53, 529 pp.

Loiacono, N. J., 1986. Hydrologic and hydrochemical characterization of the shallow ground-water system of the Great Marsh, Indiana Dunes National Lakeshore, Indiana: Proceedings of the Seventh Annual Water Resources Symposium, Indiana Water Resources Association, June 4-6, 1986, Angola, Indiana.

Schneider, A. F., 1966. Physiography. In: Natural features of Indiana, Lindsey, A. A., (Editor). Indiana Academy of Science and Indiana State Library, p. 40-56.

Shedlock, R. J., 1983. Hydrogeology of Indiana Dunes National Lakeshore--A National Park in an urban-industrial setting. Abstracts with Programs, Geological Society of America, Annual Meeting 15 (6): 685.

Shedlock, R. J. and W. E. Harkness, 1984. Shallow ground-water flow and drainage characteristics of the Brown ditch basin near the East Unit, Indiana Dunes National Lakeshore, Indiana, 1982: U. S. Geological Survey Water-Resources Investigations Report 83-4271, 37 pp.

Skougstad, M. W., M. J. Fishman, L. C. Friedman, D. E. Erdmann, and S. S. Duncan (Editors), 1979. Methods for determination of inorganic substances in water and fluvial sediments. U.S. Geological Survey Techniques of Water-Resources Investigations, Book 5, Chapter A1, 626 pp.

Thompson, T. A., 1985. Late Pleistocene lacustrine sediments of the Cowles Bog area, Indiana Dunes National Lakeshore. Abstracts with Programs, Geological Society of America 17: 329.

Toth, J., 1963. A theoretical analysis of ground-water flow in small drainage basins. Journal of Geophysical Research 68: 4795-4812.

Wilcox, D. A., Shedlock, R. J., and Hendrickson, W. H., 1986. Hydrology, water chemistry, and ecological relations in the raised mound of Cowles Bog. Journal of Ecology 74: 1103-1117.

Wood, W. W., 1976. Guidelines for collection and field analysis of ground-water samples for selected unstable constituents. U. S. Geological Survey Techniques of Water-Resources Investigations, Book 1, Chapter D2, 24 pp.

EFFECTS OF COAL FLY-ASH DISPOSAL ON WATER CHEMISTRY IN AN INTRADUNAL WETLAND AT INDIANA DUNES

Douglas A. Wilcox[1,2]
Mark A. Hardy[3]

ABSTRACT

An intradunal wetland within the Indiana Dunes National Lakeshore on the south shore of Lake Michigan was flooded for 15 years by seepage from fly-ash settling ponds located adjacent to the park. Studies were undertaken to determine the effects of the seepage on water chemistry in the flooded wetlands. These water chemistry conditions have been correlated to ongoing studies of soil contamination and secondary succession in the wetland basin following cessation of seepage. The seepage increased the concentrations of calcium, potassium, sulfate, aluminum, boron, iron, manganese, molybdenum, nickel, strontium, and zinc in ground water and surface water downgradient from the settling ponds. Chemical interactions with aquifer materials, particularly organic matter, significantly limit the transport of aluminum, iron, nickel, and zinc in this system. The organic soils of the dewatered wetland basin now contain elevated concentrations of aluminum, boron, manganese, and zinc that are potentially phytotoxic under the low pH (<4) conditions that exist. Plant growth and secondary succession were affected by the soil chemistry of the dewatered wetlands.

INTRODUCTION

The use of coal as a fuel in electrical generating facilities has increased as a result of the recognized long-term limits on petroleum-based fuel supplies. The major waste product of coal combustion is fly ash, small (0.5 - 100 um diameter) inorganic particles that are entrained in the gas stream and carried up the stack following combustion. Fly ash is composed primarily of the oxides of silica, aluminum, iron, and calcium, with smaller amounts of magnesium, titanium, sulfur, sodium, and potassium. Trace elements are also found in fly ash, either on the surface of the particle or incorporated into its matrix (Campbell et al., 1978). To prevent much of the fly ash from entering the atmosphere, electrostatic precipitators are used to collect the ash. The ash is normally transported to a landfill in dry form or sluiced to settling ponds in a water slurry. Leachates from fly ash are known to contain inorganic ions that can affect biota in aquatic receiving systems (see review by Cherry et al., 1984).

---
[1] National Park Service, Indiana Dunes National Lakeshore, 1100 N. Mineral Springs Road, Porter, Indiana 46304
[2] Present address: U. S. Fish and Wildlife Service, National Fisheries Research Center-Great Lakes, 1451 Green Road, Ann Arbor, Michigan 48105
[3] U.S. Geological Survey, 810 Bear Tavern Rd., West Trenton, New Jersey 08628

The purpose of this paper is to 1) describe the effects of fly-ash leachate on water resources, particularly wetlands, and 2) discuss the role of water chemistry studies in an interdisciplinary wetland research project.

STUDY AREA

The study area (Figure 1) is located on the Calumet Lacustrine Plain in Porter County in northwest Indiana, partly within the boundaries of Indiana Dunes National Lakeshore. The area is within a large complex of dunes deposited during a level of ancestral Lake Michigan higher than the modern lake level. Many of the lowlands within the dune complex are wetlands. The largest wetland addressed in this study has historically been called Blag Slough (Cook and Jackson, 1978). It was probably a seasonally-flooded wetland prior to disturbance of the natural hydrologic regime (Wilcox et al., 1985).

A coal-fired electrical generating station was constructed near the site in the mid-1960's, and fly-ash settling ponds were constructed adjacent to the southern edge of the wetland. The station burns coal mined primarily from Indiana and Illinois having medium to high sulfur contents (Magee et al., 1973). Leachate from the fly-ash ponds seeped into Blag Slough through a sand dike at a rate of about 7.5 million liters/day (Meyer and Tucci, 1978), causing water levels to rise and become stabilized. As a result, Blag Slough and unnamed wetlands to the north had become perennially flooded by 1967.

Seepage from the fly-ash ponds was terminated in 1980 when the ponds were drained and sealed, and the wetland basins were dry by August of that year. Low-lying areas of the Blag Slough basin have flooded seasonally in subsequent years.

METHODS

Sampling sites for water chemistry analyses were established along a transect beginning upgradient of the fly-ash settling ponds, including the settling ponds, and ending with downgradient ground and wetland surface-water sites (Figure 1). The transect is generally parallel to the direction of ground-water flow. Sites 1 and 3 were observation wells, screened in the surface aquifer, that were installed as part of the Meyer and Tucci (1978) study. The well casing at upgradient control site 1 was constructed of ABS plastic; the casing at site 3 in the sand dike was low carbon steel. Site 2 was a surface-water sampling site in the settling pond immediately adjacent to Blag Slough, and sites 4, 5, and 6 were surface-water sampling sites in the wetlands.

Most sites were sampled 12-16 times during the period September 1976 to May 1978. All surface water samples were grab samples collected 2-3 m from shore in waters that were not vertically stratified. Ground water samples were collected from the wells. Sample pH was measured in the field at the time of

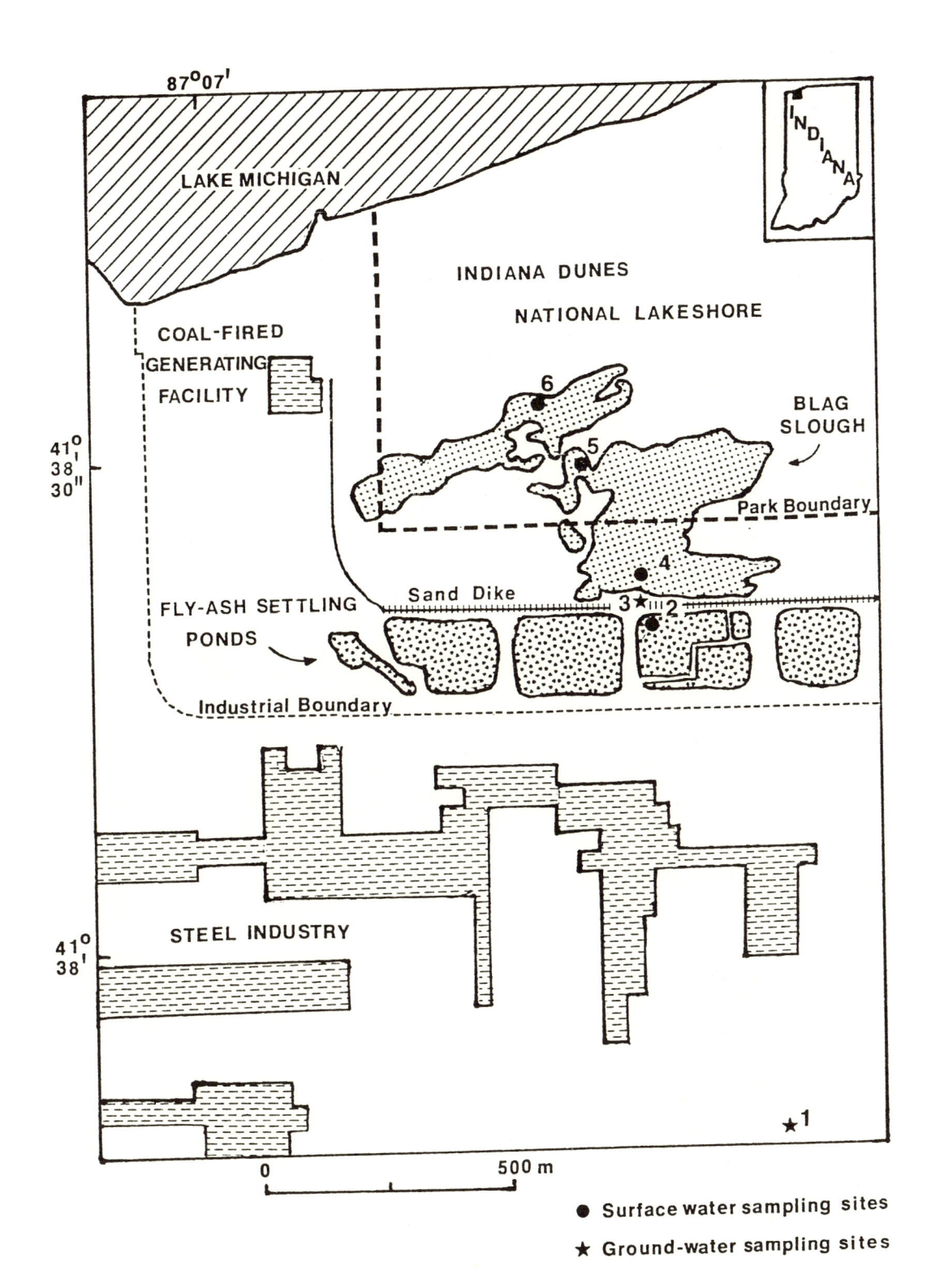

Figure 1. Site location map of Blag Slough, adjacent wetlands, coal-fired generating facility, and fly-ash settling ponds. Sampling sites and National Park Service boundaries are also shown.

sampling. All other analyses were conducted by the U.S. Geological Survey's Central Laboratory in Doraville, Georgia according to the methods of Brown et al. (1970). One dry fly-ash sample was also collected and chemically analyzed.

Log-transformed water chemistry data (except pH) from the six sampling sites were compared by one-way analysis of variance (ANOVA) and Scheffe multiple range tests. The latter procedure was selected because it is valid when sample sizes are unequal. For pH, the ANOVA was performed using actual values, which were more normally distributed than hydrogen ion concentrations. All means reported are geometric means.

RESULTS

The fly-ash sample contained high levels of leachable, sorbed calcium, silica, sodium, sulfur, aluminum and iron (Table 1). Moderate concentrations of potassium, magnesium, and boron were also present. The pH of 4.9 represents the free hydrogen concentration of the ash slurry but not the reserve (buffered) concentration.

Table 1. Chemical analysis of fly-ash sample from electrical generating station. Concentrations are in mg/kg, except for pH, and represent leachable rather than total concentrations.

Parameter	Concentration
pH	4.9
Calcium	12,000
Chloride	11
Fluoride	92
Magnesium	770
Potassium	990
Silica	150,000
Sodium	2,000
Sulfur	10,000
Aluminum	5,000
Arsenic	100
Barium	50
Beryllium	3
Boron	490
Cadmium	3
Chromium	2
Cobalt	10
Copper	30
Iron	4,600
Lead	60
Manganese	60
Mercury	0.2
Molybdenum	22
Nickel	10
Selenium	0.2
Strontium	70
Zinc	90

Site 1 upgradient of the fly-ash ponds had calcium bicarbonate waters, and all downgradient waters were calcium sulfate types. Compared to the upgradient site, calcium, potassium, sulfate, and hydrogen (pH) were significantly enriched ($\alpha = 0.05$) in and/or downgradient of the settling ponds (Table 2). Alkalinity and chloride were significantly lower in the settling pond and downgradient sites than in the upgradient control site. No statistical differences were found for fluoride, magnesium, silica, or sodium.

Table 2. Mean concentrations of major ions at sampling sites along transect. Concentrations are in mg/l, except for pH.

Indicator	1	2	3	4	5	6
pH (units)	7.5	5.7	5.9	6.7	7.6	7.5
Alkalinity as $CaCO_3$	210	2	5	22	44	35
Calcium	80	94	123	98	110	100
Magnesium	30	27	28	23	24	22
Sodium	32[1]	19	18	16	16	16
Potassium	2.0	18.1	22.2	16.3	18.9	14.8
Silica	12	6	8	6	3	2
Chloride	70[1]	10	10	11	11	10
Fluoride	0.1	0.3	0.4	0.3	0.2	0.2
Sulfate	65	398	458	342	357	338

[1]Site 1 is probably affected by road salting.

The pH of waters in the settling pond (site 2) varied widely (3.2-7.7). However, the mean values at site 2 (5.7) and site 3 (5.9) were both significantly lower than upgradient at site 1 (7.5). Although the pH in Blag Slough downgradient from settling ponds was not significantly different from site 1, alkalinity data indicate that ash-pond seepage significantly reduced the buffering capacity of water at all downgradient sites. Significant enrichment of calcium over the upgradient concentration was found only for sites 3 and 5. Potassium and sulfate, however, were enriched at all downgradient sites. Values for fluoride were greater at downgradient sites, but the differences were not significant.

Trace elements that were significantly enriched ($\alpha = 0.05$) in and/or downgradient of the fly-ash settling ponds were aluminum, boron, iron, manganese, molybdenum, nickel, strontium, and zinc (Table 3). Although enriched in the settling pond (site 2), concentrations of arsenic, barium, cadmium, cobalt, copper, and lead were not significantly enriched downgradient of the pond. No statistical differences were found for beryllium, chromium, mercury, or selenium.

Table 3. Mean concentrations of enriched trace elements at sampling sites along transect. Concentrations are in µg/l.

Indicator	\multicolumn{6}{c}{Sampling Site}
	1	2	3	4	5	6
Aluminum	22	556	431	107	59	--
Boron	104	2106	2896	2236	2248	1851
Iron	16	378	1676	355	122	111
Manganese	6	169	307	391	224	95
Molybdenum	3	53	33	87	214	--
Nickel	3	135	47	44	26	--
Strontium	123	490	478	424	483	--
Zinc	30	644	638	113	38	--

Of the trace elements enriched downgradient of the settling ponds, aluminum, iron, nickel, and zinc showed direct gradient declines in concentration from site 3 to site 6. Boron, manganese, molybdenum, and strontium did not display as clear a pattern.

DISCUSSION

Water Chemistry

The acidic nature of the fly-ash leachate from mid-west coal (Kopsick and Angino, 1981; Peffer, 1982) is of special concern in this study because the biological availability of many trace elements is generally higher at low pH values. Dissolution of trace elements from fly-ash particles is also generally greater at lower pH values (Andren et al., 1980; Mejstrik and Pospisil, 1983). Although the total quantity of boron leached from fly ash is independent of pH, the leaching is greater into acid solution (Cox et al., 1978). A major cause of acid formation is the dissolution of the sulfur oxides from fly-ash particles (Guthrie et al., 1982). Chu et al. (1978), Andren et al. (1980), and Pagenkopf and Connolly (1982) suggest that the pH of ash-sluicing water is primarily determined by the relative quantities of sulfate and alkaline earth metal oxides dissolved from the ash. The raw data from the fly-ash ponds show that low pH values were generally found in samples with lower ratios of alkaline earth metals (primarily calcium) to sulfate (Hardy, 1981). The effects of the acidic leachates at the study site are shown by the predominance of low pH and/or alkalinity data from sites downgradient of the settling ponds. Variations in pH from 3.2 - 7.7 in the ash pond are probably the result of ashes from a variety of sources having different sulfur contents. Higher pH levels at sites 4, 5, and 6 may be the result of reactions of the leachate with calcareous materials in the dune sands that comprise the surficial aquifer and dike.

Because water levels have receded and the wetlands of Blag Slough no longer support true aquatic fauna such as fish, the trace elements of major concern are those that are available for plant uptake and are known to exhibit toxic effects. The trace elements that were enriched in the leachates analyzed in this study are among those commonly reported to leach from fly ash (Cherry and Guthrie, 1977; Kopsick and Angino, 1981; Dodd, 1983; Humenick et al., 1983; Mika et al., 1985). Of these, uptake and concentration in the tissues of aquatic and wetland plants has been demonstrated for aluminum (Furr et al., 1979; Guthrie and Cherry, 1979a, 1979b), boron (Kovacs et al., 1984), manganese (Hutchinson and Czyrska, 1975; Guthrie and Cherry, 1979a, 1979b; Aulio and Salin, 1982), nickel (Aulio and Salin, 1982; Kovacs et al., 1984), and zinc (Folsom and Lee, 1981; Schierup and Larsen, 1981; Aulio and Salin, 1982).

Although aluminum, boron, manganese, nickel, and zinc are all reported to cause toxic effects in certain agricultural plants (U.S. Environmental Protection Agency, 1976; Foy et al., 1978), documentation of toxic effects of these elements on aquatic/wetland plants is scarce. Hutchinson and Czyrska (1975) found a synergistic toxic effect of cadmium and zinc and also of copper and nickel in studies with Lemna (duckweed) and Salvinia (floating fern). However, van der Werff and Pruyt (1982) found no clear evidence of zinc toxicity in four aquatic plant species, despite considerable tissue accumulations. Similar results were reported by Cherry et al. (1984) and Mika et al. (1985).

Current research stresses the importance of nutrient uptake from the sediments by aquatic plants (see review by Spence, 1982) and also indicates that trace elements leached from fly ash may accumulate on sediments rather than remaining in the water column (Guthrie and Cherry, 1979b). In fact, the pattern of declines in dissolved concentrations of aluminum, iron, nickel, and zinc in Blag Slough downgradient of the settling ponds may reflect accumulation by the sediments. Folsom and Lee (1981) further indicate that trace metals accumulated in flooded sediments may become more available to plants when the sediments dry out due to organic decomposition. Therefore, the information obtained from the study of water chemistry in Blag Slough indicates that potentially toxic materials from the fly-ash leachate may be available for plant uptake from the soils of the now-dewatered wetland.

Coupling With Other Disciplines

Work in other disciplinary fields has been conducted in conjunction with the water chemistry studies to provide a better understanding of the processes and effects of the seepage of fly-ash leachate into the Blag Slough wetlands. The hydrologic studies of Meyer and Tucci (1978) provided evidence that a ground-water mound from fly-ash pond seepage was actually responsible for the flooding of Blag Slough. Further hydrologic studies by Cohen and Shedlock (1986) documented the decay of the ground-water mound following cessation of seepage in 1980.

An interdisciplinary study of secondary succession in the Blag Slough wetlands was initiated in 1982 (Wilcox and Hiebert,

1984). The soils of the wetland basin were classified and mapped; shallow piezometers and land surveying were used to construct a water table map, a topographic map, and a contour map of the depth to the water table; major vegetation types were mapped and sampled quantitatively in successive years; and the wetland seed bank was assessed. In addition, a study of the fossil pollen record from a sediment core was undertaken (Futyma, 1986). This combination of studies assessed vegetative history and pointed out the importance of the moisture regime in contributing to the revegetation patterns observed. The extremely slow regeneration process was attributed partly to a lack of abundant viable seeds. Soil chemistry and contamination played a major role in determining which plant species were capable of surviving in the wetland following cessation of seepage. As stated previously, some of the contaminants should be present in the soil in plant-available forms. Soil chemistry data from the site were summarized by Wilcox et al. (1985) and are presented in Table 4. Zinc, manganese, boron, and aluminum were found in potentially phytotoxic concentrations in a low pH range where each is plant-available. In addition, the low soil pH may independently prevent the invasion of many common weedy species. Symptoms of boron phytotoxicity (vein-clearing) were observed in shrub species in the field and are being assessed by tissue analysis (Wilcox and Hiebert, unpublished). Bioassays utilizing soils from Blag Slough have shown reduced plant growth, morphological abberations, and poor root development (a symptom of aluminum toxicity) (Wilcox and Hiebert, 1984).

Table 4. Maximum trace element concentrations (μg/g) and pH (units) in soils from <u>Scirpus cyperinus</u> vegetation types, 1984[1].

Vegetation Type	Zn	Mn	B	Al	pH
<u>Scirpus cyperinus</u>	34	24	2.2	500	3.7
<u>Scirpus</u>/<u>Polygonum</u>	26	42	2.0	600	3.8
<u>Scirpus</u>/mixed forb mosaic	13	14	3.2	190	4.0

[1]From Wilcox et al., 1985.

The water chemistry study, which is the subject of this paper, is an integral part of the overall research project. There is an obvious relationship between the chemistry and hydrology of the seepage reaching the wetland. In addition, relationships between biological and environmental factors can be more easily explored because the water chemistry data provide the framework for understanding potential soil contamination and plant toxicity.

Clearly, correlations can be drawn between all aspects of the overall project, each discipline providing additional information to help understand the results of the others. Management implications and options, ranging from the Department of the Interior agreement with the power company to seal the fly-ash ponds to any possible amelioration attempts, must rely on as complete an understanding of all the wetland functions as is possible.

ACKNOWLEDGMENTS

The authors express their appreciation to John Tyler and Cheryl Silcox for assistance in data collection, Robert J. Shedlock and Thomas Imbrigiotta for their helpful reviews of the manuscript, and Ms. Cathie Zaharias for typing the manuscript.

LITERATURE CITED

Andren, A., M. Anderson, N. Loux, and R. Talbot, 1980. Element flow in aquatic systems surrounding coal-fired power plants. U.S. Environmental Protection Agency Report EPA 600/3-80-076. Duluth, Minnesota, 83 pp.

Aulio, K. and M. Salin, 1982. Enrichment of copper, zinc, manganese, and iron in five species of pondweeds (Potamogeton spp.). Bulletin of Environmental Contamination and Toxicology 29:320-325.

Brown, E., M. W. Skougstad, and M. J. Fishman, 1970. Methods of collection and analysis of water samples for dissolved minerals and gases. U.S. Geological Survey Techniques of Water-Resources Investigations, Book 5, Chapter A1.

Campbell, J., J. Laul, K. Neilson, and R. Smith, 1978. Separation and chemical characterization of finely-sized fly ash particles. Analytical Chemistry 50: 1032-1040.

Cherry, D. S. and R. K. Guthrie, 1977. Toxic metals in surface waters from coal ash. Water Resources Bulletin 13: 1227-1236.

Cherry, D. S., R. K. Guthrie, E. M. Davis, and R. S. Harvey, 1984. Coal ash basin effects (particulates, metals, acidic pH) upon aquatic biota: an eight year evaluation. Water Resources Bulletin 20: 535-544.

Chu, T. J., R. J. Ruane, and P. A. Krenkel, 1978. Characterization and reuse of ash pond effluents in coal-fired power plants. Journal Water Pollution Control Federation 50: 2494-2508.

Cohen, D. A. and R. J. Shedlock, 1986. Shallow ground-water flow, water levels, and quality of water, 1980-1984, Cowles Unit, Indiana Dunes National Lakeshore. U.S. Geological Survey Water-Resources Investigations Report 85-4340. 25 pp.

Cook, S. G. and R. S. Jackson, 1978. The Bailly Area of Porter County, Indiana. Robert Jackson and Associates, Evanston, Illinois, 110 pp.

Cox, J. A., G. L. Lundquist, A. Przyjazny, and C. D. Schmulbach, 1978. Leaching of boron from coal ash. Environmental Science and Technology 12: 722-723.

Dodd, D. J. R., 1983. Coal-fired power generation in Ontario - a 60 year phenomenon. Coal ash disposal in Ontario - a 1000 year legacy? Water Science and Technology 15: 65-82.

Folsom, B. L. and C. R. Lee, 1981. Zinc and cadmium uptake by the freshwater marsh plant Cyperus esculentus grown in contaminated sediments under reduced (flooded) and oxidized (upland) disposal conditions. Journal of Plant Nutrition 3: 233-244.

Foy, C. D., R. L. Chaney, and M. C. White, 1978. The physiology of metal toxicity in plants. Annual Reviews of Plant Physiology 29: 511-566.

Furr, A. K., T. F. Parkinson, W. D. Youngs, C.O. Berg, W.H. Gutenmann, I. S. Pakkala, and D. J. Lisk, 1979. Elemental content of aquatic organisms inhabiting a pond contaminated with coal fly ash. New York Fish and Game Journal 26: 154-161.

Futyma, R. P., 1986. Fossil pollen stratigraphy of Blag Slough, Indiana Dunes National Lakeshore. National Park Service, Porter, Indiana, 24 pp.

Guthrie, R. K. and D. S. Cherry, 1979a. Trophic level accumulation of heavy metals in a coal ash basin drainage system. Water Resources Bulletin 15: 244-248.

Guthrie, R. K. and D. S. Cherry, 1979b. The uptake of chemical elements from coal ash and settling basin effluent by primary producers. I. Relative concentrations in predominant plants. The Science of the Total Environment 12: 217-222.

Guthrie, R. K., E. M. Davis, D. S. Cherry, and J. R. Walton, 1982. Impact of coal ash from electric power production on changes in water quality. Water Resources Bulletin 18: 135-138.

Hardy, M. A., 1981. Effects of coal fly-ash disposal of water quality in and around the Indiana Dunes National Lakeshore, Indiana. U.S. Geological Survey Water-Resources Investigations Report 81-16. 64 pp.

Humenick, M. J., M. Lang, and K. F. Jackson, 1983. Leaching characteristics of lignite ash. Journal Water Pollution Control Federation 55: 310-316.

Hutchinson, T. C. and H. Czyrska, 1975. Heavey metal toxicity and synergism to floating aquatic weeds. Internationale Vereinigung fur theoretische und angewandte Limnologie 19: 2102-2111.

Kopsick, D. A. and E. E. Angino, 1981. Effect of leachate solutions from fly and bottom ash on groundwater quality. Journal of Hydrology 54:341-356.

Kovacs, M., I. Nyary, and L. Toth, 1984. The microelement content of some submerged and floating aquatic plants. Acta Botanica Hungarica 30: 173-185.

Magee, E. M., H. J. Hall, and G. M. Varga, 1973. Potential pollutants in fossil fuels. U.S. Environmental Protection Agency Report EPA-R2-73-249, National Technical Information Service, PB 225 039, 292 pp.

Mejstrik, V. and J. Pospisil, 1983. Sorption of Cd, Cu, Cr, Zn, and Pb in fly ash and environmental impacts. Ekologia (CSSR) 2: 295-301.

Meyer, W. and P. Tucci, 1978. Effects of seepage from fly-ash settling ponds and construction dewatering on ground-water levels in the Cowles Unit, Indiana Dunes National Lakeshore, Indiana. U.S. Geological Survey Water-Resources Investigations Report 78-138. 95 pp.

Mika, J. S., K. A. Frost, W. A. Feder, and C. J. Puccia, 1985. The impact of land-applied incinerator ash residue on a freshwater wetland plant community. Environmental Pollution (Series A) 38: 339-360.

Pagenkopf, G. K. and J. M. Connolly, 1982. Retention of boron by coal ash. Environmental Science and Technology 16: 609-613.

Peffer, J. R., 1982. Fly ash disposal in a limestone quarry. Ground Water 20: 267-273.

Schierup, H.-H. and V. J. Larsen, 1981. Macrophyte cycling of zinc, copper, lead, and cadmium in the littoral zone of a polluted and a non-polluted lake. I. Availability, uptake, and translocation of heavy metals in Phragmites australis (Cav). Trin. Aquatic Botany 11: 197-210.

Spence, D. H. N., 1982. The zonation of plants in freshwater lakes. Advances in Ecological Research 12: 37-124.

U.S. Environmental Protection Agency, 1976. Quality criteria for water. U.S. Environmental Protection Agency Report EPA-440/9-76-023, 501 pp.

van der Werff, M. and M. J. Pruyt, 1982. Long-term effects of heavy metals on aquatic plants. Chemosphere 11: 727-739.

Wilcox, D. A. and R. D. Hiebert, 1984. Secondary succession and related environmental factors in a formerly flooded wetland following termination of fly-ash pond seepage. Bulletin of the Ecological Society of America 65(2): 66-67.

Wilcox, D. A., N. B. Pavlovic, and M. L. Mueggler, 1985. Selected ecological characteristics of Scirpus cyperinus and its role as an invader of disturbed wetlands. Wetlands 5: 87-99.

PREHISTORIC AND HISTORIC TRENDS IN ACIDITY OF KETTLE PONDS IN THE CAPE COD NATIONAL SEASHORE: IMPLICATIONS FOR MANAGEMENT

Marjorie Green Winkler[1]

ABSTRACT

The reconstruction of pH from diatom assemblages in cores from Duck Pond in the Cape Cod National Seashore provides evidence that some outwash kettle ponds are naturally-acid ecosystems. Duck Pond has been acid for 12,000 years with a mean reconstructed pH for the entire period of 5.2, standard deviation ± 0.3 (standard error of the predictive equation ± 0.45). This research illuminates several problems in management of the Cape Cod freshwater kettle ponds caused by concern about impacts from acid precipitation in the region. Liming of ponds is used to counter effects from acid precipitation, and liming is being advocated and carried out on a national level by the Electric Power Research Institute through an organization called Living Lakes. On the Cape, application of lime would significantly change naturally-acid ponds, some of which are presently also impacted by increased erosion and leachates from wastewater disposal. Naturally-acid ponds and the organisms that have evolved within them, are, therefore, endangered ecosystems, and management decisions involving these complex biologic communities will have major effects on preservation of these ponds. Interdisciplinary research is needed to define environmental change and paleoecological techniques are needed to provide a long-term perspective, but interjurisdictional decision-making in response to environmental problems is also crucial. Natural areas within the parks (such as the Great Ponds in the Cape Cod National Seashore) are governed by both state and federal agencies that operate under different management directives, and these ecosystems are also affected by land-use policies made by local governments outside of the parks. Park and local communities must base management on site-specific research findings, although national policies are needed to solve complex environmental problems.

INTRODUCTION

Recent increases in toxic and acidic substances in precipitation are known to have caused major problems in wetlands in sensitive regions of North America (Gorham et al., 1984). Areas that are most affected by acid precipitation are New England, the Adirondacks in New York State, and parts of the southeastern and western states (Likens et al., 1979). Although recent lake damage is indicated in some studies, the cause of within-lake changes such as the disappearance of fish is not clear

---
[1]Center for Climatic Research, Institute for Environmental Studies, University of Wisconsin-Madison, 1225 West Dayton Street, Madison, Wisconsin 53706

(Kahl et al., 1985; Ford, 1986). In the United States, some studies of sensitive regions are presently being conducted, but few long-term lake pH histories have been completed (Battarbee, 1984; Winkler, 1985a; Charles and Norton, 1986). Because of this circumstance, another problem is surfacing. Natural resource agencies and groups, both governmental and private, are advocating "treatment" of acid lakes "before it is too late." Since the damages of acid precipitation have been rightly publicized, for most policy makers in sensitive areas, the decision to "treat" acid lakes with applications of agricultural limestone is usually not contested, and debate focuses on when and how much, not whether.

A study of the pH history of Duck Pond in the Cape Cod National Seashore suggests that some lakes are naturally acid ecosystems (Winkler, 1985a; 1988). "Treatment" of naturally-acid lakes with limestone would change these ecosystems. These findings emphasize the need for long-term pH histories of individual lake systems and a classification that separates culturally-acidified lakes from naturally-acid ones. Naturally-acid lakes may have evolved buffering systems--possibly composed of organic acids and/or metal compounds--that react to natural fluctuations in acidity. The results of this study also emphasize the complexity of environmental issues and the need for inter-jurisdictional communication, so that ecologically sound, site-specific management decisions can be made.

Cape Cod, Massachusetts, may not be alone in facing this problem, for lake water chemistry is being measured in many areas and is used to identify modern low-pH lakes, but the long-term pH history of these lakes is not known. Other places in North America where naturally-acid lakes would be located include igneous bedrock, crystalline outwash, and coastal sand deposits in the southeastern United States, northcentral Florida, the northern Midwest, the Adirondacks and northern New England, and Canada.

STUDY AREA

The Cape Cod National Seashore was dedicated in 1961 to protect a unique and diverse landscape in the easternmost part of Massachusetts (Figure 1). Included in this park are seacoasts, dunes, salt marshes, freshwater marshes and bogs, cedar swamps, and freshwater kettle ponds. The mosaic of the present vegetation on the Outer Cape reflects an ocean-tempered climate, strong salt- and sand-laden winds from numerous storms, and the soils and topographic diversity of the underlying pitted outwash plain. A forest of Pinus rigida (pitch pine) and scrub and other oaks (Quercus sp.) is the dominant modern vegetation on the dry ridges and slopes with a groundcover of ericads [Vaccinium sp. (blueberries), Arctostaphylos uvi-ursi (bearberry)], Comptonia peregrina (sweetfern), and Myrica pensylvanica (bayberry), while more mesophytic trees [Carya sp. (hickory), Fagus grandifolia (beech), Acer rubrum (red maple), Nyssa sylvatica (sour gum), and Chamaecyparis thyoides (Atlantic white cedar)] grow in the wetter hollows.

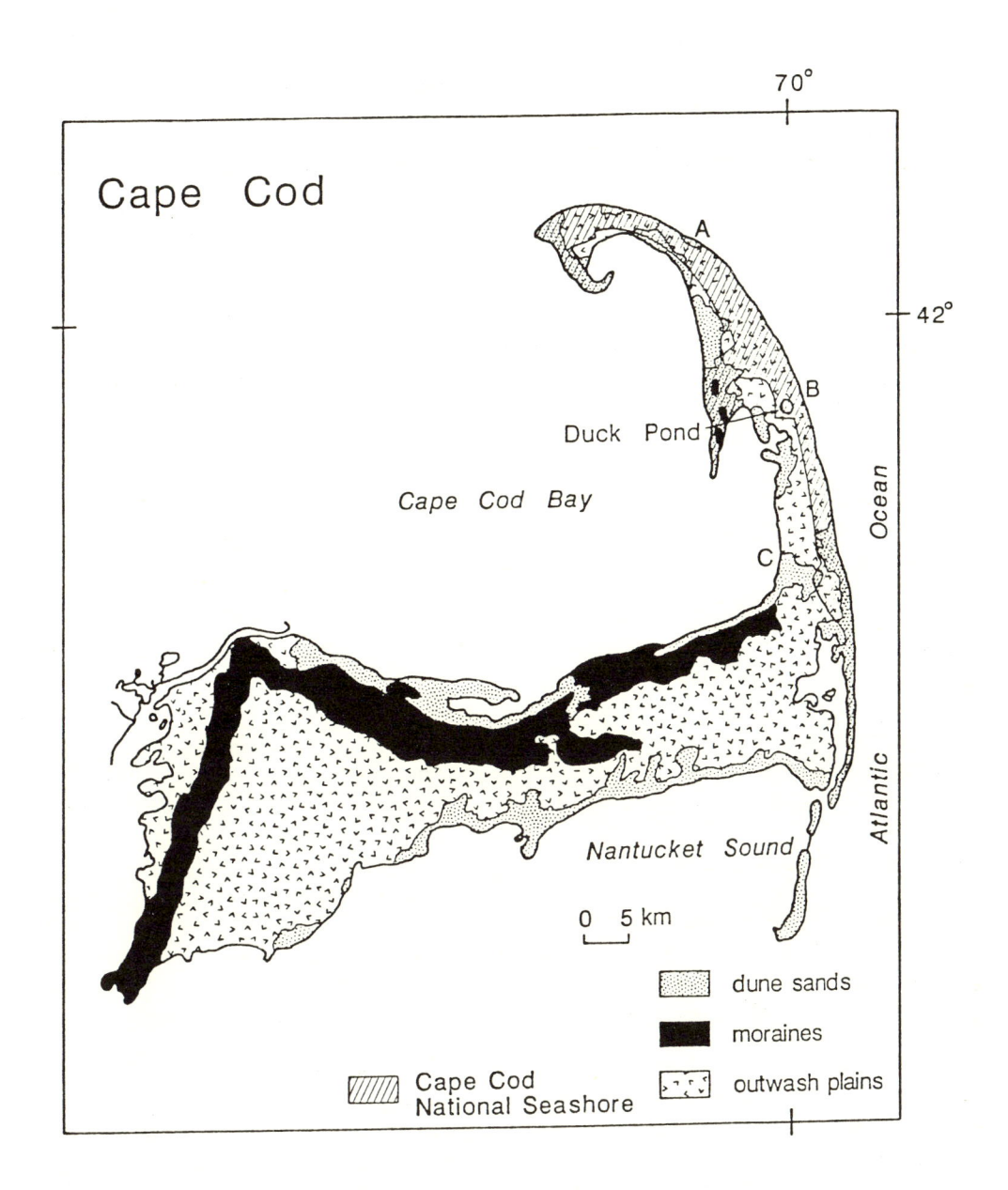

Figure 1. Map of the edaphic and geologic characteristics of Cape Cod, Massachusetts. A, B, and C mark the areas of the Outer Cape that have ponds included in the diatom-pH study (Winkler, 1985a). Duck Pond is labeled at B. The boundary of the Cape Cod National Seashore is also demarcated.

Duck Pond (41°56'N, 70°00'W) is 5.1 ha in area and is 2.5m above mean sea level. It has a maximum depth of 18.5m and a mean depth of 5.9m (Soukup, 1977) and is one of the many oligotrophic kettle ponds within the pitted crystalline outwash plains of the Outer Cape (Winkler, 1985a,b).

METHODS

Sediment cores were obtained from the deepest part of Duck Pond with a modified Livingstone piston corer and with a Davis-Doyle corer that provides shorter cores containing the sediment-water interface (see Winkler, 1985a,b). The cores were 4.28m and 1m long, respectively. After subsampling, the lake sediment was processed for pollen, charcoal, and diatoms (see Winkler, 1985a,b). Diatom identification was accomplished using floras listed in Winkler (1985a) and the diatom stratigraphy was then divided into pH-associated groups based on the Hustedt (1939) diatom-pH group classifications:

Acidobionts (Acb): diatom taxa which occur at pH lower than 5.5,
Acidophils (acp): diatoms found primarily below pH 7,
Circumneutral (indifferent) (Ind): diatoms occurring at pH 7,
Alkaliphils (alk): diatom taxa found primarily above pH 7,
Alkalibionts (Akb): diatoms which occur only above pH 7.

Pearson product-moment linear correlation matrices were calculated for the chemical, morphometric, and biological variables, and Statgraphics was used to generate and test a multiple linear regression equation from the diatom and chemical data.

PREHISTORIC AND HISTORIC TRENDS IN ACIDITY OF KETTLE PONDS IN THE CAPE COD NATIONAL SEASHORE: SUMMARY OF RESULTS

The relationship between modern diatom assemblages and the modern lakewater chemistry of the ponds in the Cape Cod National Seashore in south-eastern Massachusetts (Winkler, 1985a; 1987), was used to reconstruct the 12,000-year pH history of Duck Pond (Winkler, 1985a; 1988).

The kettle ponds in the Cape Cod National Seashore are set in non-calcareous, crystalline, outwash sands and have low specific conductance, low alkalinity, low productivity, and generally clear water. These factors make them sensitive to acid and toxic deposition because they receive most of their nutrients from the atmosphere. The 15 Outer Cape ponds chosen for study range in pH from 4.3 to 7.5 and are broadly representative of the more than 180 ponds on the Cape (with a pH range of 4.1 to 8.3) (WMP, 1976). The ponds can be divided into three groups: ponds with a pH below 5, those with moderate pH values from 5 to 6.8, and ponds with some tidal influence and pH values higher than 6.8.

Examination of pondwater pH measurements taken intermittently over the past few years and monthly for the past two years shows that the pH increased about 0.3 pH units in several of the ponds from low values in September, 1975 and April, 1982 and has then remained relatively stable (Figure 2) (monthly pH measurements of the pondwater since February 1983 were provided by the Acid Rain Monitoring Project [ARMP], see Winkler, 1985a). Precipitation pH, on the other hand, has varied greatly during the past 4 years, ranging between very low (less than pH 3.5) to moderate (more than pH 5.6) (NADP data, see Winkler, 1985a). Precipitation on the Outer Cape contains abundant seawater ions as well as high nitrates and sulfates during the lowest pH events. Although there is no apparent seasonality to low pH precipitation events, increased summer use of the ponds (Soukup, 1977) provides alkaline

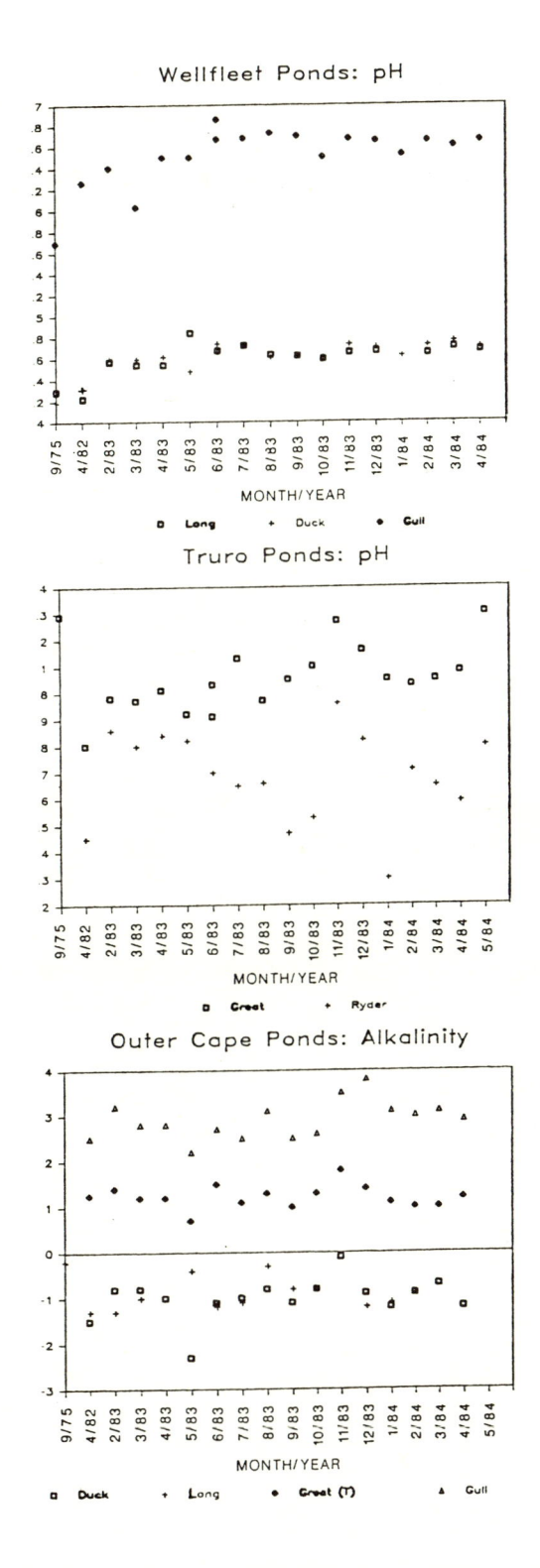

Figure 2. pH and alkalinity measurements from 9/75 to 5/84 for ponds in Wellfleet and Truro, Massachusetts, located between A and B on Figure 1, within the Cape Cod National Seashore boundary. The pH was measured potentiometrically on pond surface water collected monthly since February, 1983 and intermittently before that time. Alkalinity was measured potentiometrically by the EPA method for low-alkaline lakes.

inputs either directly into the ponds via bathing and washing (use of shampoo and soap) or indirectly (through discharge of household water into the shallow groundwater aquifer of the Outer Cape) that balance some of the acidity in summer rainfall. On-site sewage disposal systems may also release water through the coarse outwash sands into the aquifer or ponds.

Diatoms in the modern sediments from the Outer Cape ponds were analyzed and divided into pH-related groups (Figure 3A). Correlation analysis suggests that each pH-related diatom group responds to different characteristics of the ponds (Winkler, 1985a). Acidobiontic diatoms are negatively correlated with alkalinity while acidophilic diatoms are negatively correlated with pH and productivity. The alkaliphilic diatoms are positively correlated with pH and productivity and circumneutral diatoms are positively correlated with the percent of total planktonic diatoms and negatively correlated with the percent of total acid diatoms (the acidobiontic plus the acidophilic diatoms for each level). The alkalibiontic diatoms are correlated with specific conductance and seawater-associated ions and reflect the coastal environment of the Cape. Of course there may be non-linear relationships between some of the variables as well.

Changes in pH over the past 12,000 years in Duck Pond, S. Wellfleet, were reconstructed by using a pH-diatom transfer function (Winkler, 1985a; 1988). The transfer function equation is:

pH = 2.47 + 0.017 (% acidobionts) + 0.023 (% acidophils) + 0.041 (% indifferent) + 0.046 (% alkaliphils) + 0.349 (% alkalibionts)

and has a standard error of + 0.45. The reconstructed pH suggests that the pond has been acid for its entire history with a mean pH of 5.2, S.D. + 0.3 (Figure 3B). The highest pH in the pond (a pH of 6) was during a period of increased windiness and erosion [inferred from pollen and sediment changes indicative of an open *Picea* (spruce) - *Hudsonia* (poverty grass) parkland (Winkler, 1985b)] during late-glacial time. Although the diatom evidence indicates that Duck Pond did become more acid recently (with a mean pH of about 4.9, S.D. + 0.1 for the past 150 years), the pH has varied both up and down throughout the Holocene, and Duck Pond has had acidity as low as the present at other times in the past 12,000 years. It must be noted, however, that the downcore reconstructed pH values remain within the error of the regression equation (S.E. + 0.45) and primarily reflect a long-term acid ecosystem.

The relationship between the pollen and charcoal stratigraphy from Duck Pond (Winkler, 1985b) and the diatom stratigraphy (Winkler, 1988) was examined. There is significant correlation between charcoal influx and alkaline *Fragilaria* diatoms in the pre-European-settlement core and between percent charcoal and the percent total acid diatoms in the post-European-settlement core.

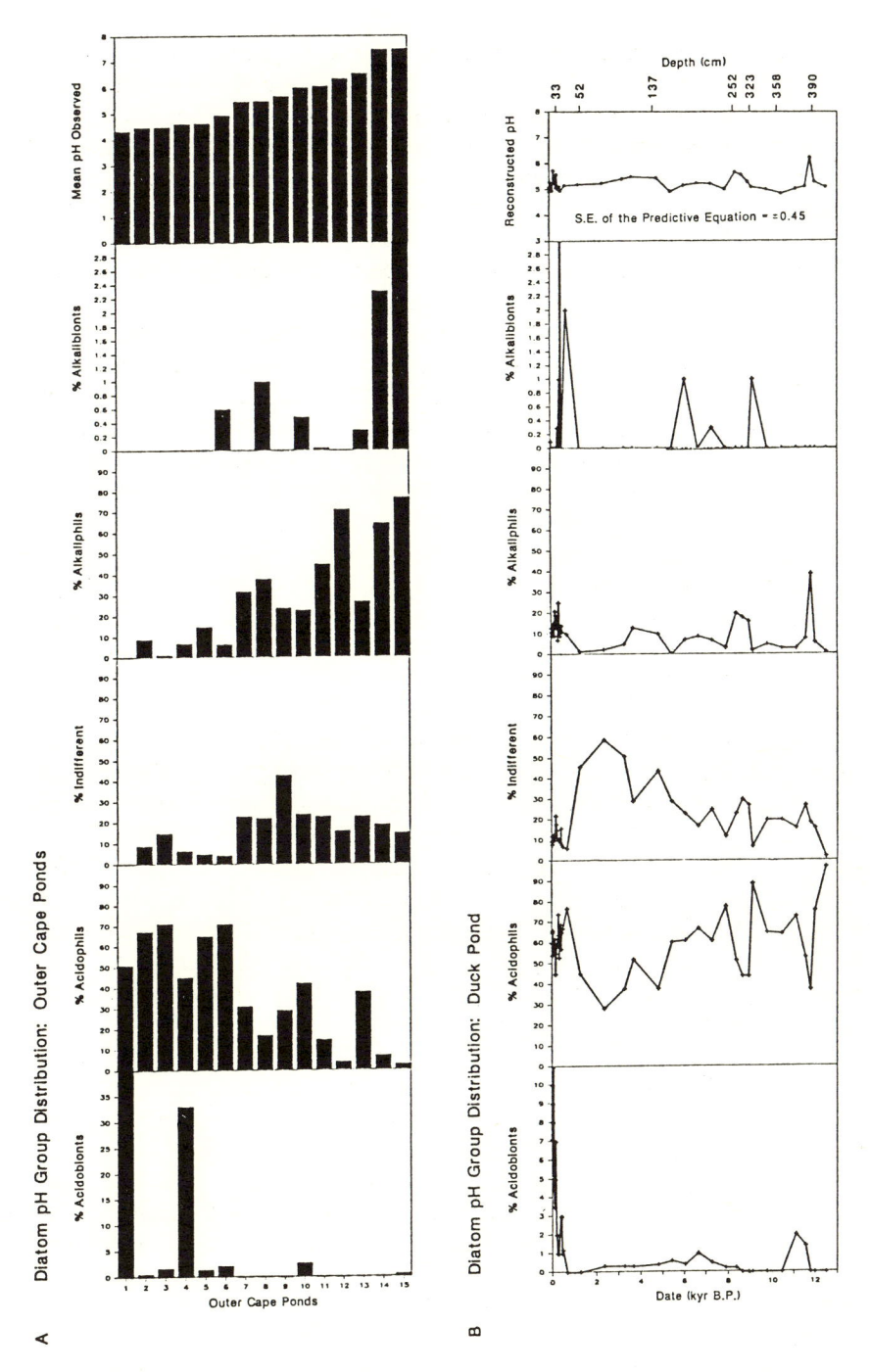

Figure 3. A. Outer Cape Ponds modern diatom-pH group distribution (follows Hustedt, 1939, pH classification, see Methods) and recent mean pH values for each pond (see Figure 2 caption). Note the scale change for the % Acidobiontic and % Alkalibiontic diatoms.

B. Duck Pond diatom-pH group distribution plotted on a time scale based on radiocarbon dates (see Winkler, 1985b). ^{14}C dates were available for all the depths noted on the right-hand scale, except at 33 cm, which is dated by an increase in ragweed pollen indicating European settlement. The reconstructed pH values for each level are also presented. Note the scale change for the % Acidobiontic and % Alkalibiontic diatoms.

Charcoal influx and percent charcoal are significantly correlated with each other, but influx takes into account the variations in the depositional environment within the lake. These depositional changes were large enough during the late-glacial and the Holocene to obscure a linear relationship between percent charcoal and the other variables. However, the relationship between charcoal influx and the alkaline *Fragilaria* diatoms is clear and significant. After European settlement, though, percent charcoal shows a linear relationship with some of the variables probably because there was little morphometric or depositional variation throughout this relatively short period of time in the history of the pond (330 years). The charcoal results may reflect the fact that wood charcoal and ash deposited in runoff from local forest fires is generally alkaline while windborne charcoal residue, the gas and soot from wood or fossil fuel combustion, is quite acid. Although fires near the pond could contribute substantially to pH changes in this low conductance, low alkalinity pond, other factors such as changes in vegetation, lake levels, and climate (especially temperature, precipitation, and atmospheric CO_2 variations) may have contributed to the pH fluctuations in Duck Pond throughout the Holocene.

A negative correlation between the diploxylon *Pinus* (pitch pine) species and the reconstructed pH suggests that increased amounts of pitch pine growing near the pond may have lowered the pH of the pond, either through acidic substances released upon decomposition, or through changes in the rate of evapotranspiration or CO_2 utilization in the watershed. The linear correlation between haploxylon *Pinus* (white pine) and the alkaliphilic diatoms, suggest that white pine has an opposite effect from pitch pine on the pH of a pond. It is possible that white pine has different compounds in its wood and needles from pitch pine. This is not surprising when the different modern distribution of these pines is examined (Fowells, 1965; Little, 1971) and the importance of tree physiology to climatic adaptation is considered. White pine has a markedly decreased adaptation to fire and salt spray compared to pitch pine. It is therefore evident that changes in vegetation caused by changes in climate and fire frequency also affected the pH history of the pond.

MANAGEMENT IMPLICATIONS OF RESEARCH RESULTS

The reconstruction of pH from diatom assemblages in cores from Duck Pond provides evidence that some of the outwash kettle ponds in the Seashore are long-term (thousands of years) naturally-acid ecosystems. The fact that Duck Pond has been acid for 12,000 years raises the following issues:

1). These results underscore the need for increasing concern with the widespread practice of liming of acid ponds to counteract damage from acid precipitation. Although liming or other treatments may be indicated in cases where anthropogenic degradation is documented, application of lime to naturally-acid kettle ponds degrades these ecosystems and eliminates the organisms that have evolved over millenia within them. Since the pH of many lakes in sensitive areas are presently being monitored, the knowledge of the long-term pH history of lakes or sets of lakes that are found to be acid is crucial to making ecologically-sound management decisions. This is true for lakes within the national parks (as in the Cape Cod National Seashore) and also outside of them. Currently, treatment with lime is being implemented on Cape Cod by the Commonwealth of Massachusetts Fish and Wildlife Division, which oversees the Great Ponds (ponds larger than 10 acres) of the state. Furthermore, the liming of ponds is being advocated and carried out by a Washington, D. C.-based group called "Living Lakes," which was formed by the Electric Power Research Institute. The availability and eagerness of "Living Lakes" to carry out liming within a community, decreases discussion and debate on the liming issue. The Cape Cod kettle ponds are also being impacted by increased development that is accompanied by high erosion rates and increasingly contaminated deposits from many sources. Therefore, naturally-acid ponds are becoming endangered entities as concern about acid precipitation damage to wetlands increases.

2) The results from this study further emphasize the need for interdisciplinary research in order to define the reactions of complex ecologic systems to complex environmental stresses. Stratigraphic analysis of sediment cores provides the needed historical perspective upon which to base ecologically-sound management decisions. These studies integrate modern pond chemistry and biology with precipitation chemistry, atmospheric dynamics, and geologic, pollen, charcoal, and diatom evidence of past changes. Diatoms are sensitive to changes in acidification and eutrophication. The study of diatom evidence of environmental change in wetlands of the Cape Cod National Seashore can suggest management programs for the freshwater kettle ponds of the Seashore and can assist in estimating what activities outside of the park (such as increased pumping of freshwater to supply new development, increased fertilizer use with a very shallow, porous aquifer, increased groundwater contamination from increased traffic and more permanent development) work against continued preservation of within-park fragile ecosystems and, again, consider remedies if necessary. Ongoing monitoring of ground water, lake water, and lake sediments is needed to provide both chemical and biological (diatom, etc.) indications of changes in the freshwater wetlands and determine the direction and the rate of changes. In addition, mineral and ion analysis is needed of modern sediments from the lakes, downcore sediments from some of the ponds, ground water from wells near the ponds, and fog aerosols.

3). These results, furthermore, point out the importance of cross-jurisdictional involvement in environmental decision making, because: a). activities outside of the park affect preservation of ecosystems within the park, and b). several agencies, both state and federal, are charged with management and preservation of the Seashore ponds --all with different management priorities.

Just as complex environmental problems, such as the issue of acid and toxic deposition, can only be illuminated by interdisciplinary research, it is increasingly evident that interdisciplinary, cross-jurisdictional decision-making is just as important to effective management and must be used to mitigate complex environmental impacts. Exploration of avenues of intergovernmental cooperation is vital because cross-jurisdiction of park assets leads to at-odds management decisions. Policy decisions based on complex information should be anticipated by setting up an interagency commission that would be able to communicate concerns and would have the power to guide important decisions. Emphasis should be placed on regional, community, or group solutions to park problems (Soukup, 1977) not only because land-use policies outside of the park boundaries affect water quantity and water quality of the ponds within the park (directly affecting their recreational use and preservation), but also because similar problems confront local communities outside of the parks. A biography of the individual park, the charter, and the extent of the interjurisdictional aspects of the various park assets should be compiled and should include statements of management intent from each jurisdiction outside of the National Park Service. This document would provide a basis for dialogue with other agencies. It would also permit prediction of management conflicts in the future and enable early settlement of policy problems. It may also be used to guide adjacent communities in making ecologically-sound conservation decisions.

CONCLUSIONS

Actions by different units of government impinge on the environmental and ecological concerns of other jurisdictional units. In this example, in the Cape Cod National Seashore, the Commonwealth of Massachusetts oversees management of the "Great Ponds of the State" (those over 10 acres in area). Thus, state agencies make policy that affects the management of the ponds in the Seashore. The Fish and Wildlife Division of the Commonwealth of Massachusetts wishes to lime acidic Cape ponds, but recent research (Winkler, 1985b) indicates that some ponds have a long, 12,000-year history of relatively low pH. Treatment with lime will change the ecology of these ponds and therefore state policy conflicts with the policy of preservation established by the Seashore. These conflicts are hard to resolve because of the different management directions of different agencies, the time it takes to modify old management decisions (i.e., liming as an overall policy), and the time it takes to develop policies based on site-specific research. Furthermore, liming of ponds is being advocated and carried out free of charge by Living Lakes, Inc., a non-profit group formed by the Electric Power Research Institute and based in Washington, D. C. Obviously, the Living Lakes organization has yet another management objective and does not consider the results of site-specific research. Site-specific, cooperative, management decisions must be made in order to preserve natural resources, although effective environmental policy concerning complex impacts from complex problems, such as acid and toxic deposition, must be made at the national level.

ACKNOWLEDGMENTS

I have had the invaluable help and support of numerous people from the Cape Cod National Seashore, the North Atlantic Regional Office of the National Park Service, and the Center for Climatic Research of the University of Wisconsin-Madison. I would especially like to thank Dr. M. A. Soukup for scientific and financial support and J. Portnoy for logistic help and scientific discussions. B. Richards drafted the figures. This research was funded by the Office of Scientific Studies, North Atlantic Regional Office, National Park Service, Order No. PX1600-3-0518 and by NSF grants ATM 82-19079 and ATM 84-12958, to J. E. Kutzbach, Center for Climatic Research, University of Wisconsin-Madison.

LITERATURE CITED

Battarbee, R. W., 1984. Diatom analysis and the acidification of lakes. Philosophical Transactions of the Royal Society of London B 305: 451-447.

Charles, D. F. and S. A. Norton, 1986. Paleolimnological evidence for trends in atmospheric deposition of acids and metals. In: Acid Deposition: Long Term Trends. National Research Council Committee on Monitoring and Assessment of Trends in Acid Deposition. National Academy Press, Washington, D. C., pp. 335-506.

Ford, J., 1986. The recent history of a naturally acidic lake (Cone Pond, N. H.). In: Diatoms and Lake Acidity, J.P. Smol, R. W. Battarbee, R. B. Davis, and J. Merilainen (Editors). Dr. W. Junk Publishers, Dordrecht.

Fowells, H. A., 1965. Silvics of Forest Trees of the United States. Agriculture Handbook No. 271, USDA Forest Service, Washington, D. C., 762 pp.

Gorham, E., F. B. Martin, and J. T. Litzau, 1984. Acid rain: ionic correlations in the Eastern United States, 1980-1981. Science 225: 407-409.

Hustedt, F., 1939. Systematische und okologische Untersuchungen uber die Diatomeen-Flora von Java, Bali, und Sumatra nach dem Material der Deutschen Limnologischen Sunda-Expedition III. Die okologischen Factorin und ihr Einfluss auf die Diatomeenflora. Archiv fur Hydrobiologie Supplementen 16: 274-394.

Kahl, J. S., J. L. Andersen, and S. A. Norton, 1985. Water resource baseline data and assessment of impacts from acidic precipitation, Acadia National Park, Maine. Technical Report #16. National Park Service, North Atlantic Region Water Resources Program.

Likens, G. E., R. F. Wright, J. N. Galloway, and T. J. Butler, 1979. Acid Rain. Scientific American 241: 43-51.

Little, Jr., E. L., 1971. Atlas of United States Trees, Vol. 1. Conifers and important hardwoods. USDA Forest Service Misc. Pub. 1146. Washington, D. C.

Soukup, M. A., 1977. Limnology and the Management of the Freshwater Ponds of Cape Cod National Seashore. Report No. 35. Cooperative Research Unit, University of Massachusetts-National Park Service. 73 pp.

Winkler, M. G., 1985a. Diatom evidence of environmental changes in wetlands: Cape Cod National Seashore. Final Report to the North Atlantic Regional Office of the National Park Service, Boston, Massachusetts. 120 pp.

Winkler, M. G., 1985b. A 12,000-year history of vegetation and climate for Cape Cod, Massachusetts. Quaternary Research 23: 301-312.

Winkler, M. G., 1988. Diatom stratigraphy and pH reconstruction of a 12,000-year old kettle pond in the Cape Cod National Seashore. Ecology 69: (in press).

WMP, 1976. Final Report Water Quality Assessment Cape Cod Ponds prepared by the Environmental Management Institute for the Area-wide Wastewater Management Program. Cape Cod Planning and Economic Development Commission, Barnstable Co., Massachusetts. Vols. I and II, 269 pp.

IMPLICATIONS OF WETLAND SEED BANK RESEARCH: A REVIEW OF GREAT BASIN AND PRAIRIE MARSH STUDIES[1]

Roger L. Pederson[2] and Loren M. Smith[3]

ABSTRACT

The objective of this paper was to review the seed bank research from Great Basin and prairie marshes relative to methodology, seed bank characteristics, and implications of seed bank research. Seed bank density estimates are extremely variable relative to species and to total germinable seedling present. Most studies should stratify sampling because seed densities are usually different by elevation and/or vegetation type. Research is needed to determine optimal sample shape to minimize variation among samples. Methodology differences and the heterogeneity of wetland seed banks make statistical comparisons among studies difficult.

The floristic composition of seed banks varied in relation to factors including elevation, vegetation type, soil depth, soil moisture, and salinity. Highest seed densities were generally found along shoreline zones in the denser vegetation types. Fewer germinable seeds were associated with high soil salinities. Several species present as seeds in the soil were not present in existing vegetation.

Seed bank studies can be used to provide information on plant life-histories, the potential flora, the historical vegetation, and the nature of seed dispersal. This data can be used to: 1) predict and interpret vegetation change in response to management procedures (e.g., water level changes, irrigation, discing, fire), and 2) design and plan wetland restorations. Several examples of wetland seed bank research are examined to illustrate the utility of this approach.

INTRODUCTION

In wetlands and other plant communities, seed banks play an important role in maintaining floristic and genetic diversity. The size and composition of the seed bank, the relative input from different species (seed rain), and the pattern of seed distribution reflect not only the nature of contemporary vegetation and its surroundings but the history of the environment as well (Major and Pyott, 1966; Harper, 1977; Baskin and Baskin, 1978; Levin and Wilson, 1978).

---
[1] Paper No. 31 of the Marsh Ecology Research Program, a joint project of the Delta Waterfowl and Wetlands Research Station and Ducks Unlimited Canada.
[2] Delta Waterfowl and Wetlands Research Station, R. R. 1, Portage la Prairie, Manitoba, Canada R1N 3A1
[3] Department of Range and Wildlife Management, Texas Tech University, Box 4169, Lubbock, Texas 79409

In this paper, we will review seed bank research of Great Basin and prairie wetlands and summarize the information with respect to: 1) methodology, 2) seed bank characteristics, and 3) theoretical and practical implications of this type of investigation. Although we have emphasized seed bank work in Great Basin and prairie marshes (because of their relatively similar environmental conditions), generalizations from other wetland seed bank research will also be discussed. For specific details on seed bank studies in lakes and other types of wetlands, the reader is directed to Moore and Wein (1977), Leck and Graveline (1979), Keddy and Reznicek (1982), Haag (1983), Nicholson and Keddy (1983), Hopkins and Parker (1984), and Parker and Leck (1985).

SEED BANK RESEARCH METHODS

Community and Species Composition

Seed bank studies in Great Basin and prairie wetlands have emphasized the "seedling emergence" method (Roberts, 1981) in which field soil samples are: 1) placed directly into shallow containers or spread in a thin layer over sterilized soil, 2) placed in a greenhouse or outdoor shelter, 3) kept moist or shallowly-flooded, and 4) monitored by identifying and recording seedlings that emerge (van der Valk and Davis, 1976a; van der Valk and Davis, 1978; Pederson, 1981; Smith and Kadlec, 1983). The objective is to ensure that as many of the viable seeds as possible germinate and produce seedlings.

The seedling emergence technique has advantages over physically sieving seeds because only viable seeds that produce seedlings are recorded. Identification of seedlings is quicker and easier than identification of seeds. Identifying monocot seedlings that fail to flower can be a problem (Smith and Kadlec, 1985a). Identification of some species, however, can potentially be solved by growing seedling sets from known seed sources (Keddy and Ellis, 1985), as well as by using histological characteristics of vegetative parts (Best et al., 1971; Metcalfe, 1971; Korschgen, 1980, 1983). Although some species may still be impossible to identify even with histological techniques and known seed sources (e.g., *Typha* spp.), these shortcomings are generally considered minor when evaluating the total community.

Disadvantages of wetland seedling emergence studies relate to the inability to detect buried seed populations of some aquatic plants. Seeds of some wetland species may be inviable or may require different germination conditions than those provided in a greenhouse. Also, differential seed production and longevity among species may cause investigators to overemphasize the importance of certain species (on the basis of buried seed populations) in past vegetation history (see van der Valk and Davis, 1979). In addition, if the study objective is to mimic what occurs in the field, greenhouse germination conditions (soil and water salinity, light, moisture, temperature) will usually be different than those of field situations (Smith and Kadlec, 1983, 1985b). These differences, as well as regional variations in soil and flora, hamper comparisons among studies from diverse areas. To clarify the interpretation of variation among studies, accurate measurements of seedling emergence conditions should be maintained for samples (see Smith and Kadlec, 1983). Light and photoperiod should approximate ambient conditions in the region, and samples should be observed for the entire growing season (seeds of various species may germinate at different times in the season--see Harris and Marshall, 1963; Thompson and Grime, 1979; Roberts, 1986).

Experimental Design

Typical design questions that arise when initiating a seed bank study include: 1) sample shape, 2) potential stratification, and 3) sample number considerations. Unfortunately, the topic has received little study, especially in wetlands. Sample shape, which will influence variability among samples, probably has received the least amount of study. Sample shape in most studies was probably selected on the basis of what was convenient or what previous studies have used (e.g., Ekman dredge or soil-coring devices), with little concern over what might be best in terms of minimizing variance (see Keddy and Reznicek, 1985, for surface area per sample in various studies). For marsh sediments, soil sampling to a depth of 4-5 cm is commonly selected. Given the logistical constraints in obtaining sediment samples, a fixed set of surface sediment sizes and shapes should be tested to determine optimal surface size and shape per sample in order to adequately sample the community and keep variation to a minimum.

Stratification of sampling within a particular wetland, in most cases, will improve seedling density estimates. For example, Pederson (1981) and Smith and Kadlec (1983, 1985a) noted that seed densities varied significantly by elevation and vegetation, respectively, in prairie and Great Basin. This is most likely the effect of seed production and resultant seed dispersal patterns along past and present environmental gradients. Thus, sampling should account for obvious or perceived gradients (e.g., elevation, soil type, etc.).

Even within a stratified site, most seed bank samples are highly variable (standard deviations approaching or exceeding density means). The number of samples required to obtain density estimates within a certain percentage of the population mean has received little study in wetlands. Champness and Morris (1948) noted that, for grasslands, 200 soil samples (each 25 cm^2 in size) would be required to obtain buried-seed density estimates within 10% of the mean for most common species. Whipple (1978) concluded that most upland seed bank studies contained too few samples to get seed density estimates within 10% of the population mean, and most studies of buried seed floras should be viewed as estimates of species presence rather than estimates of seed densities. This is also true for most wetland studies. If the objective is to merely characterize wetland species composition, a species-area curve should be constructed to determine the number of samples needed until no new species are found (Harper, 1977).

Although a large number of samples are necessary to get estimates within a certain percentage of the buried-seed population mean, fewer samples may be required to discern differences among areas or vegetation types (Pederson, 1981; Smith and Kadlec, 1983, 1985a). Smith and Kadlec (1985a) found that variance estimates did not decrease appreciably as sample size increased from 10 to 25 within the same marsh vegetation type. However, with few samples, even though variances may be the same as for more samples, the ability to detect differences among vegetation types will be low and a Type II error may occur.

Because of high variance estimates associated with seed bank samples, parametric tests of differences in seedling density among zones or vegetation types are often difficult without data transformation (Zar, 1984) due to violation of parametric assumptions (e.g., non-normal distributions). Rank transformation of data with subsequent parametric analysis is another option (Conover and Iman, 1981). There are also many non-parametric test alternatives for those studies not meeting parametric assumptions (Hollander and Wolfe, 1973).

SEED BANK CHARACTERISTICS OF GREAT BASIN AND PRAIRIE WETLANDS

Seed Bank Development

Although Keddy and Reznicek (1985) generalized that buried seed densities seem to increase from prairie marshes to freshwater tidal wetlands to lakeshores, methodology differences and the extreme heterogeneity of wetland seed banks make statistical comparisons among wetlands difficult (van der Valk and Davis, 1976a; van der Valk and Davis, 1979; Roberts, 1981). Table 1 summarizes data from seed bank studies of marshes in Iowa (semi-permanently flooded, freshwater), Manitoba (permanently flooded, oligosaline), and Utah (seasonally-flooded, oligosaline-mixosaline). The greatest species diversity was found in seed banks from the Iowa wetlands, whereas the Utah seed banks were the least diverse. All seed banks were dominated by seed populations of mudflat annuals and flood-intolerant, perennial species (e.g., *Scirpus validus*). Also, with the exception of soil samples from the Iowa wetlands, seed populations in the substrate from open-water sites were consistently lower than seed populations present in soil from emergent stands. Generalizations from these data indicate that the nature of the seed bank in a particular prairie wetland is a function of its hydrology, sediment and water characteristics, and existing vegetation.

Numerous studies have documented the dramatic impact of water level changes on vegetation in Great Basin and prairie wetlands (Harris and Marshall, 1963; Weller and Spatcher, 1965; Stewart and Kantrud, 1972; van der Valk and Davis, 1978). Vegetation recolonization, seed production, and seed dispersal during low-water

Table 1. Number of species, most abundant genera, and mean number of seedlings/m^2 that emerged from soil samples (moist-soil treatment[a]) taken from emergent vegetation and open-water sites.

Geographic location and wetland type[b]	Number of sample sites	Number of species in seed bank	Location of sample site	Seedlings/ m^2	Most abundant species[c] in seed bank (in descending order)	Seedlings/ m^2
Iowa, palustrine emergent wetland, semi-permanently flooded, fresh (van der Valk and Davis, 1978)	24	40	emergent veg. open-water	4073 3549	Scirpus validus Bidens cernua Polygonum lapathifolium Typha spp. Rorippa islandica	4178 2018 770 713 538
Manitoba, lacustrine emergent wetland, permanently flooded, oligosaline (Pederson, 1981)	318	34	emergent veg. open-water	4582 298	Scirpus validus Typha spp. Chenopodium rubrum Ranunculus sceleratus Scolochloa festucacea	5463 1955 1338 201 277
Utah, palustrine emergent wetland, seasonally-flooded, oligosaline-mixosaline (Smith and Kedlec, 1983)	150	24	emergent veg. open-water	2932 70	Polypogon monspeliensis Typha spp. Chenopodium rubrum Rumex crispus Distichlis spicata	3216 1388 938 404 340

[a]Soil samples were watered every day.
[b]After Cowardin et al., 1979.
[c]Bidens, Polygonum, Rorippa, Chenopodium, Ranunculus, Polypogon, and Rumex are mudflat species; Scirpus, Distichlis, and Scolochloa are perennial emergents intolerant of prolonged flooding; Typha is a perennial emergent tolerant of prolonged flooding.

periods are major factors determining seed bank input from mudflat annuals and flood-intolerant perennial species (van der Valk and Davis, 1979; Keddy and Reznicek, 1982; Pederson and van der Valk, 1984; Smith and Kadlec, 1985a). Conversely, periods of high water allow for seed input from submersed aquatics (Haag, 1983; Smith and Kadlec, 1985a) and perennial emergents that endure flooding (Pederson and van der Valk, 1984).

In brackish prairie wetlands, soil salinity levels during low-water periods may restrict vegetation colonization and seed production (Christiansen and Low, 1970; Lieffers and Shay, 1981), and seed banks from these wetlands express this constraint (Smith and Kadlec, 1983).

In wetlands where perennial emergent vegetation is not eliminated by periodic drought, flooding, or other disturbances, sizable seed banks may not develop because: 1) the rapid regrowth from residual vegetation restricts seedling recruitment of many annuals and short-lived perennials, and 2) the dominant perennial species may produce few viable seeds (van der Valk and Davis, 1979; Hopkins and Parker, 1984; Pederson and van der Valk, 1984; Smith and Kadlec, 1985a, b).

Seed Dispersal

In prairie pothole marshes in Iowa, van der Valk and Davis (1976a) found considerable among-marsh variation in the composition and densities of seed banks; however, they were unable to detect within-marsh differences (presumably because seeds had been dispersed to all areas within a marsh). In contrast, considerable within-marsh variation (e.g., between open-water and emergent vegetation sites) in seed bank composition exists in other types of wetlands (Table 1). High seed densities in soil samples from emergent stands in littoral zones of marshes in Utah and Manitoba reflected shoreline seed accumulations caused by water movement (Pederson, 1981; Smith and Kadlec, 1983; Pederson and van der Valk, 1984). Hall et al. (1946) and Smith and Kadlec (1985a) have shown that seed movement during and after disturbance (water level changes, fire, etc.) could be as important as buried seed reserves in determining the composition and pattern of subsequent vegetation recruitment. Although wind and water dispersal of aquatic plant seeds have been shown to be significant in forest openings (Wagner, 1965) and irrigation canals (Eggington and Robbins, 1920; Kelley and Bruns, 1975), little information is available quantifying seed movement during disturbance events in wetlands.

Seed Distribution in Soil Profiles

Van der Valk and Davis (1978) categorized vegetation composition changes in prairie glacial wetlands as occurring in four phases: the dry marsh, regenerating marsh, degenerating marsh, and lake marsh. Because a wetland's seed bank receives varied seed input during each of the phases, the different layers of a marsh's substrate should contain a record of past vegetational change. By examining the nature of seed distribution in a soil profile, it should be possible to reconstruct the vegetational history in the same manner that paleoecological studies have used pollen and mollusc profiles to determine regional vegetation history (e.g., Watts and Winter, 1966; Watts and Bright, 1968).

The vertical distribution of buried seeds (determined from seedling assay methods) under Typha glauca communities in prairie marshes in Iowa (van der Valk and Davis, 1979) and Manitoba (Pederson and van der Valk, 1984) is illustrated in Table 2. Van der Valk and Davis (1979) examined buried seed populations in alternate 5 cm soil layers to a depth of 35 cm, whereas Pederson and van der Valk (1984) examined buried seed populations in each 2 cm layer from the soil surface to a depth of 8 cm. Generalizations about seed populations in wetland soil profiles indicate large populations of germinable seeds (e.g., 47,200 seeds/m^2 at the 20-25 cm soil depth in the Iowa study--Table 2) can occur deep within a soil profile. Barton (1961), Harper (1977), and Roberts (1981) have all commented on how seeds of annual and aquatic plants remain viable longer than seeds of other groups of plants. For the Iowa data (Table 2), the presence of mudflat species throughout the profile and the seed population changes of Scirpus validus and Typha spp. indicate there have been periodic drawdowns and vegetation changes at that site. The Manitoba data (Table 2) suggest that the site was once much drier (large seed accumulations of annual species in the 4-8 cm soil layers), then became wetter (diminished seed input from annuals and increased seed input from Zannichellia palustris), and the dominant vegetation had shifted from Scirpus to Typha.

Methodology problems associated with using seedling-assay methods to reconstruct the past vegetation history of a wetland include the inability to detect seeds of many dominant emergents (e.g., Carex spp., Phragmites communis, Sparganium spp., Scirpus acutus, Scirpus fluviatilis--van der Valk and Davis, 1979; Pederson, 1981). This problem may be partially solved by mechanically sieving soil samples after the seedling-assay treatment and identifying seeds (see Watts and Winter, 1966; Roberts, 1981). Additionally, certain perennial species (e.g., Typha, Scirpus validus) are prolific seed producers and, because of seed dispersal, these species may be over-represented in soil samples collected in areas in which they were not dominant (see Wagner, 1965; van der Valk and Davis, 1979). This problem can be partly clarified by examining seed populations in soil cores from several different locations in the wetland (e.g., replicate core sites per vegetation cover and elevation interval) and determining seed densities that represent background levels resulting from seed dispersal (van der Valk and Davis, 1979). Finally, although changes in seed populations throughout a soil core can indicate periods of time when the wetland was dry or flooded, pinpointing exact dates is not possible because little is known of sediment accumulation rates in wetlands. However, as van der Valk and Davis (1979) pointed out, despite the many interpretational and technical problems of using seedling-assay methods to reconstruct

Table 2. Mean number of seeds and propagules/m^2 found at different soil depths of _Typha glauca_ communities in prairie wetlands in Iowa and Manitoba. Samples were processed under moist-soil and submersed conditions in the greenhouse.

Geographic Location and Wetland Type[a]	Species[b]	Soil Core Section (Depth from Surface)				
		0-5 cm	10-15 cm	20-25 cm	30-35 cm	
Iowa, palustrine emergent wetland, semi-permanently flooded, fresh (van der Valk and Davis, 1979[c])	_Chara_ sp.	1200	831	369	462	
	Cyperus odoratus	1105	910	845	65	
	Polygonum lapathifolium	2405	3365	1073	553	
	Rorippa islandica	553	1463	1983	293	
	Rumex maritimus	1235	1463	1593	715	
	Scirpus validus	13358	47353	38870	3055	
	Typha spp.	1939	1268	1385	554	
Total Number of Seeds[d]		25182	56289	47200	7075	
		0-2 cm	2-4 cm	4-6 cm	6-8 cm	
Manitoba, lacustrine emergent wetland, permanently flooded, oligosaline (Pederson and van der Valk, 1984[e])	_Chenopodium rubrum_	25	0	475	775	
	Ranunculus sceleratus	0	250	1650	300	
	Rumex maritimus	0	0	950	25	
	Scirpus maritimus	12	122	585	95	
	Scirpus validus	606	6125	29250	4762	
	Typha spp.	625	125	63	0	
	Zannichellia palustris	1775	812	737	0	
Total Number of Seeds[d]		3068	7446	33710	5957	

[a]After Cowardin et al., 1979.
[b]Only the most abundant species are shown. _Cyperus_, _Chenopodium_, _Polygonum_, _Ranunculus_, _Rorippa_, and _Rumex_ are mudflat species; _Scirpus_ spp. are perennial emergents intolerant of prolonged flooding; _Typha_ is a perennial emergent tolerant of prolonged flooding; _Chara_ and _Zannichellia_ are submersed aquatics.
[c]Adapted from Table II.
[d]Totals reflect counts of all species found in samples, not just the species listed in the Table.
[e]Adapted from Table 2.

wetland vegetation history, these problems are similar in nature and scope to those palynologists faced in the past (see Faegri and Iversen, 1975).

IMPLICATIONS OF SEED BANK RESEARCH

Seed bank research in wetlands has both theoretical and practical implications in the study of plant population biology (e.g., vegetation dynamics and life-history strategies) and wetland management (e.g., classification, restoration, and vegetation management).

Vegetation Dynamics and Life-history Strategies

Plant communities in wetlands often occur as distinct zones or bands of vegetation that follow basin contours (Cowardin et al., 1979). Although some introductory ecology texts still infer that vegetation zonation in wetlands represents succession, in most cases, the two phenomena are unrelated (van der Valk and Davis, 1978; van der Valk, 1981; Keddy and Reznicek, 1985). Vegetation zonation in wetlands is better viewed as the response of different wetland species to environmental gradients characteristic of those environments (Sculthorpe, 1967; Hutchinson, 1975; Davis and Brinson, 1980; Spence, 1982).

Past studies of vegetation dynamics in prairie wetlands have emphasized adult plant distribution in relation to one or more environmental gradients (e.g., water quality, water level fluctuation, disturbance--Walker and Coupland, 1968, 1970; Walker and Wehrhahn, 1971; Stewart and Kantrud, 1972; Millar, 1973; van der Valk and Davis, 1976b). Field investigations or experiments with adult plants or ramets have assumed that adult stage interactions are largely responsible for vegetation zonation. During dry conditions in wetlands, however, the vegetation can change dramatically as species intolerant of drying die and are replaced by species emerging from reserves of buried seeds (Kadlec, 1962; Harris and Marshall, 1963; van der Valk and Davis, 1978, 1979; Cooke, 1980). The degree of seed bank influence on vegetation dynamics of a particular wetland site depends on: 1) the amount of residual vegetation capable of growing vegetatively, 2) the composition of the seed bank, 3) germination and seedling emergence conditions present at the site, and 4) seed input into the site.

Van der Valk (1980, 1981) used seed bank information (potential flora) plus field studies (current flora) to develop a qualitative model for predicting wetland succession. Aquatic plant species were classified into 12 basic life-history types on the basis of: 1) life span, 2) propagule longevity, and 3) propagule establishment requirements. The wetland environment was viewed as a sieve that permits establishment of only certain life-history types at any given time. Although a starting point for future studies on wetland vegetation dynamics, van der Valk's model is qualitative (it predicts presence or absence of a species and does not predict relative abundance). Smith and Kadlec (1985b) found van der Valk's model predictions relatively satisfactory in predicting species composition after a fire in a Utah marsh. However, to improve the quantitative capability of the model, Smith and Kadlec (1985b) recommended incorporating into the model: 1) variables to account for the overriding effects of residual, perennial vegetation; and 2) continuous variables (e.g., salinity, light, soil temperature, and soil moisture gradients), which would influence both the type and amount of seedling expression from the seed bank (see Hall et al., 1946; Harris and Marshall, 1963; Connelly, 1979; Thompson and Grime, 1979; Simpson et al., 1983; Keddy and Ellis, 1985; Parker and Leck, 1985; Galinato and van der Valk, 1986; van der Valk, 1986).

Wetland Classification, Restoration, and Management

The latest wetland classification system (Cowardin et al., 1979) defines wetlands as having one or more of the following attributes: 1) at least periodically, the land supports predominantly hydrophytes, 2) the substrate is predominantly undrained hydric soil, and 3) the substrate is nonsoil and saturated with water or covered by shallow water at some time during the growing season of each year. While this system is useful for defining wetlands that presently exist, these criteria will not delimit former wetlands whose hydrology has been altered by man (e.g., riverine marshes that have been affected by altered hydrologic flows or marshes that have been drained for agricultural purposes). Because large populations of buried aquatic seeds are known to exist in shoreline zones (Nicholson and Keddy, 1983; Smith and Kadlec, 1983; Pederson and van der Valk, 1984), detection of buried seed populations of certain aquatic plants (e.g., species with water-dispersed achenes and grains) could be used to define former contours of altered wetland basins or riverine marshes.

Additionally, by examining the seed bank of a drained wetland, insights about the potential flora, past vegetation history, and hydrology could be used to plan and manage restorations (e.g., Dunn and Best, 1983). Seed bank information can be used in predicting not only the future species composition of the reclaimed wetland (van der Valk, 1980; Pederson and van der Valk, 1984), but also in determining the proper sequence and timing of water-level manipulations during restoration (Hall et al., 1946; Harris and Marshall, 1963; Kadlec and Wentz, 1974; Garbisch, 1977; Landin, 1978; Herner and Co., 1980). Erwin and Best (1985) demonstrated the utility of using natural wetland seed banks (soil from a wetland burrow area) as a mulch-topsoil addition to a surface-mined, reclaimed wetland in Florida. The latent wetland flora (seeds and propagules) contained within the mulch permitted a more diverse and rapid revegetation of a reclaimed wetland area than did overburdened topsoil (depauperate in aquatic plant seeds) spread on other reclaimed sites (Erwin and Best, 1985).

Finally, characterizing seedling expression from wetland seed banks in relation to seasonality (Hall et al., 1946; Harris and Marshall, 1963; Connelly, 1979) and environmental gradients (Smith and Kadlec, 1983; Kadlec and Smith, 1984; Pederson and van der Valk, 1984; Knighton, 1985) has led to many marsh management techniques and schedules that favor certain plant species over others. These techniques enable marsh managers to manipulate the vegetation composition and structure in order to enhance wetlands for wildlife populations and may be especially useful in controlling problem plant species (Spencer and Bowes, 1985).

LITERATURE CITED

Barton, L. V., 1961. Seed Preservation and Longevity. Leonard Hill, London, 216 pp.

Baskin, J. M. and C. C. Baskin, 1978. The seed bank in a population of an endemic plant species and its ecological significance. Biological Conservation 14:125-130.

Best, K. F., J. Looman, and J. B. Campbell, 1971. Prairie grasses identified and described by vegetative characters. Canadian Department of Agriculture Publication 1413, 239 pp.

Champness, S. S. and K. Morris, 1948. The population of buried viable seeds in relation to contrasting pasture and soil types. Journal of Ecology 36:149-173.

Christiansen, J. E. and J. B. Low, 1970. Water requirements of waterfowl marshlands in northern Utah. Utah Division Fish and Game Publication 69-12, Salt Lake City, Utah, 108 pp.

Connelly, D. P., 1979. Propagation of selected native marsh plants in the San Joaquin Valley. Wildlife Management Leaflet No. 15. California Department of Fish and Game, Sacramento, California, 13 pp.

Conover, W. J. and R. L. Iman, 1981. Rank transformation as a bridge between parametric and nonparametric statistics. American Statistician 35:124-133.

Cooke, G. D., 1980. Lake level drawdown as a macrophyte control technique. Water Resources Bulletin 16:317-322.

Cowardin, L. M., V. Carter, F. C. Golet, and E. T. LaRoe, 1979. Classification of wetlands and deep water habitats of the United States. FWS/OBS-79/31. Office of Biological Services, Fish and Wildlife Service, U.S.D.I., Washington, D.C., 103 pp.

Davis, G. J. and M. M. Brinson, 1980. Responses of submerged vascular plant communities to environmental change. FWS/OBS-79/33. Office of Biological Services, Fish and Wildlife Service, U.S.D.I., Washington, D.C., 70 pp.

Dunn, W. J. and G. R. Best, 1983. Seed bank survey of some Florida marshes and the role of seed banks in marsh reclamation. In: Enhancing Ecological Succession. Proceedings of the National Symposium on Surface Mining, Hydrology, Sedimentology, and Reclamation. Office of Continuing Education, University of Kentucky, Lexington, Kentucky, pp. 365-370.

Eggington, G. E. and W. W. Robbins, 1920. Irrigation water as a factor in the dissemination of weed seed. Colorado Agricultural Experiment Station Bulletin 253.

Erwin, K. L. and G. R. Best, 1985. Marsh community development in a central Florida phosphate surface-mined reclaimed wetland. Wetlands 5:155-166.

Faegri, K. and J. Iversen, 1975. Textbook of Pollen Analysis. 3rd Edition. Hafner, New York, 295 pp.

Galinato, M. I. and A. G. van der Valk, 1986. Seed germination traits of annuals and emergents recruited during drawdowns in the Delta Marsh, Manitoba, Canada. Aquatic Botany 26:89-102.

Garbisch, E. W., Jr., 1977. Recent and planned marsh establishment work throughout the contiguous United States--a survey and basic guidelines. Dredged Material Research Program, Contract Report D-77-3. U.S.A.E. Waterways Experiment Station, Vicksburg, Mississippi, 167 pp.

Haag, R. W., 1983. Emergence of seedlings of aquatic macrophytes from lake sediments. Canadian Journal of Botany 61:148-156.

Hall, T. F., W. T. Penfound, and A. D. Hess, 1946. Water level relationships of plants in the Tennessee Valley with particular reference to malaria control. Journal of Tennessee Academy of Science 21:18-59.

Harper, J. L., 1977. Population Biology of Plants. Academic Press, London and New York, 892 pp.

Harris, S. W. and W. H. Marshall, 1963. Ecology of water level manipulations of a northern marsh. Ecology 44:331-343.

Herner and Co., 1980. Publication and index retrieval system. Technical Report DS-78-23. U.S.A.E. Waterways Experiment Station, Vicksburg, Mississippi, 187 pp.

Hollander, M. and D. A. Wolfe, 1973. Nonparametric Statistical Methods. Wiley, New York, 503 pp.

Hopkins, D. R. and V. T. Parker, 1984. A study of the seed bank of a salt marsh in northern San Francisco Bay. American Journal of Botany 71:348-355.

Hutchinson, G. E., 1975. A Treatise on Limnology. Volume III. Limnological Botany. John Wiley and Sons, New York, 660 pp.

Kadlec, J. A., 1962. Effects of a drawdown on a waterfowl impoundment. Ecology 43:267-281.

Kadlec, J. A. and L. M. Smith, 1984. Marsh plant establishment on newly flooded salt flats. Wildlife Society Bulletin 12:388-394.

Kadlec, J. A. and W. A. Wentz, 1974. State-of-the-art survey and evaluation of marsh plant establishment techniques: induced and natural. Contract Report D-74-9. Volume I. Report of Research. U.S.A.E. Waterways Experiment Station, Vicksburg, Mississippi, 231 pp.

Keddy, P. A. and T. H. Ellis, 1985. Seedling recruitment of 11 wetland plant species along a water level gradient: shared or distinct responses? Canadian Journal of Botany 63:1876-1879.

Keddy, P. A. and A. A. Reznicek, 1982. The role of seed banks in the persistence of Ontario's coastal plain flora. American Journal of Botany 69:13-22.

Keddy, P. A. and A. A. Reznicek, 1985. Vegetation dynamics, buried seeds, and water level fluctuations on the shorelines of the Great Lakes. In: Coastal Wetlands, H. H. Prince and F.M. D'Itri (Editors). Lewis Publishing, Inc., Chelsea, Michigan, pp 33-58.

Kelley, A. D. and V. F. Bruns, 1975. Dissemination of weed seeds by irrigation water. Weed Science 23:486-493.

Knighton, M. D., 1985. Vegetation management in water impoundments: water-level control. In: Water Impoundments for Wildlife: A Habitat Management Workshop, General Technical Report NC-100, M. D. Knighton (Compiler). U.S.D.A. Forest Service, North Central Forest Experiment Station, St. Paul, Minnesota, 136 pp.

Korschgen, L. J., 1980. Histological characteristics of plant leaf epidermis and related structures as an aid in food habits studies (monocotyledons). Final Reports: Study No. 32, Job No. 1, Federal Aid Project No. W-13-R-34. Missouri Department of Conservation, Columbia, Missouri.

Korschgen, L. J., 1983. Histological characteristics of plant leaf epidermis and related structures as an aid in food habits studies (dicotyledons). Final Reports: Study No. 32, Job No. 1, Federal Aid Project No. W-13-R-37. Missouri Department of Conservation, Columbia, Missouri.

Landin, M. C., 1978. Annotated tables of vegetation growing on dredged material throughout the United States. Dredged Material Research Program, Misc. Paper D-78-7. U.S.A.E. Waterways Experiment Station, Vicksburg, Mississippi, 155 pp.

Leck, M. A. and K. J. Graveline, 1979. The seed bank of a freshwater tidal marsh. American Journal of Botany 66:1006-1015.

Levin, D. A. and J. B. Wilson, 1978. The genetic implications of ecological adaptations in plants. In: Structure and Functioning of Plant Populations, A. H. J. Freysen and J. W. Woldendorp (Editors). North Holland, Amsterdam, Oxford, New York, pp. 75-100.

Lieffers, V. J. and J.M. Shay, 1981. The effects of water level on growth and reproduction of *Scirpus maritimus* var. *paludosus* on the Canadian prairies. Canadian Journal of Botany 59:118-121.

Major, J. and W. T. Pyott, 1966. Buried viable seeds in two California bunchgrass sites and their bearing on the definition of a flora. Vegetatio 13:253-282.

Metcalfe, C. R., 1971. Anatomy of the Monocotyledons. V. Cyperaceae. The Clarendon Press, Oxford, 597 pp.

Millar, J. B., 1973. Vegetation changes in shallow marsh wetlands under improving moisture regime. Canadian Journal of Botany 51:1443-1457.

Moore, J. M. and R. W. Wein, 1977. Viable seed populations by soil depth and potential site recolonization after disturbance. Canadian Journal of Botany 55:2408-2412.

Nicholson, A. and P. A. Keddy, 1983. The depth profile of a shoreline seed bank in Matchedash Lake, Ontario. Canadian Journal of Botany 61:3293-3296.

Parker, V. T. and M. A. Leck, 1985. Relationships of seed banks to plant distribution patterns in a freshwater tidal wetland. American Journal of Botany 72:161-174.

Pederson, R. L., 1981. Seed bank characteristics of the Delta Marsh, Manitoba: applications for wetland management. In: Selected Proceedings of the Midwest Conference on Wetland Values and Management, B. Richardson (Editor). St. Paul, Minnesota, pp. 61-69.

Pederson, R. L. and A. G. van der Valk, 1984. Vegetation change and seed banks in marshes: ecological and management implications. Transactions of the North American Wildlife and Natural Resources Conference 49:271-280.

Roberts, H. A., 1981. Seed banks in soils. Advances in Applied Biology 6:1-55.

Roberts, H. A., 1986. Seed persistence in soil and seasonal emergence in plant species from different habitats. Journal of Applied Ecology 23:639-656.

Sculthorpe, C. D., 1967. The Biology of Aquatic Vascular Plants. Edward Arnold, London, 843 pp.

Simpson, R. L., R. E. Good, M. A. Leck, and D. F. Whigham, 1983. The ecology of freshwater tidal wetlands. Bioscience 33:255-259.

Smith, L. M. and J. A. Kadlec, 1983. Seed banks and their role during a drawdown of a North American marsh. Journal of Applied Ecology 20:673-684.

Smith, L. M. and J. A. Kadlec, 1985a. The effects of disturbance on marsh seed banks. Canadian Journal of Botany 63:2133-2137.

Smith, L. M. and J. A. Kadlec, 1985b. Predictions of vegetation change following fire in a Great Salt Lake marsh. Aquatic Botany 21:43-51.

Spence, D. H. N., 1982. The zonation of plants in freshwater lakes. Advances in Ecological Research 12:37-125

Spencer, W. and G. Bowes, 1985. Limnophila and Hygrophila: a review and physiological assessment of their weed potential in Florida. Journal of Aquatic Plant Management 23:7-16.

Stewart, R. E. and H. A. Kantrud, 1972. Vegetation of prairie potholes, North Dakota, in relation to quality of water and other environmental factors. U.S. Geological Survey Professional Paper 585-D, U.S. Gov't. Printing Office, Washington, D.C., 36 pp.

Thompson, K. and J. P. Grime, 1979. Seasonal variation in the seed banks of herbaceous species in ten contrasting habitats. Journal of Ecology 67:893-921.

van der Valk, A. G., 1980. Succession in temperate North American wetlands. In: Wetlands: Ecology and Management, B. Gopal, R. E. Turner, R. G. Wetzel, and D. F. Whigham (Editors). International Scientific Publications, Jaipur, India, pp. 169-179.

van der Valk, A. G., 1981. Succession in wetlands: a Gleasonian approach. Ecology 62:688-696.

van der Valk, A. G., 1986. The impact of litter and annual plants on recruitment from the seed bank of a lacustrine wetland. Aquatic Botany 24:13-26.

van der Valk, A. G. and C. B. Davis, 1976a. The seed banks of prairie glacial marshes. Canadian Journal of Botany 54:1832-1838.

van der Valk, A. G. and C. B. Davis, 1976b. Changes in composition, structure, and production of plant communities along a perturbed wetland coenocline. Vegetatio 32:87-96.

van der Valk, A. G. and C. B. Davis, 1978. The role of seed banks in the vegetation dynamics of prairie glacial marshes. Ecology 59:322-335.

van der Valk, A. G. and C. B. Davis, 1979. A reconstruction of the recent vegetational history of a prairie marsh, Eagle Lake, Iowa, from its seed bank. Aquatic Botany 6:29-51.

Wagner, R. H., 1965. The annual seed rain of adventive herbs in a radiation damaged forest. Ecology 46:517-520.

Walker, B. H. and R. T. Coupland, 1968. An analysis of vegetation-environment relationships in Saskatchewan sloughs. Canadian Journal of Botany 46:509-522.

Walker, B. H. and R. T. Coupland, 1970. Herbaceous wetland vegetation in the aspen grove and grassland regions of Saskatchewan. Canadian Journal of Botany 48:1861-1878.

Walker, B. H. and C. F. Wehrhahn, 1971. Relationships between derived vegetation gradients and measured environmental variables in Saskatchewan wetlands. Ecology 52:85-95.

Watts, W. A. and R. C. Bright, 1968. Pollen, seed, and mollusk analysis of a sediment core from Pickerel Lake, northeastern South Dakota. Bulletin Geological Society of America 79:855-876.

Watts, W. A. and T. C. Winter, 1966. Plant macrofossils from Kirchner Marsh, Minnesota--a paleoecological study. Bulletin Geological Society of America 77:1339-1359.

Weller, M. W. and C. E. Spatcher, 1965. Role of habitat in the distribution and abundance of marsh birds. Special Report No. 43. Iowa State University Agriculture and Home Economics Experiment Station, Ames, Iowa, 31 pp.

Whipple, S. A., 1978. The relationship of buried, germinating seeds to vegetation in an old-growth Colorado subalpine forest. Canadian Journal of Botany 56:1505-1509.

Zar, J. H., 1984. Biostatistical Analysis. Prentice-Hall, Inc., New Jersey, 718 pp.

SURFACE HYDROLOGY AND PLANT COMMUNITIES OF CORKSCREW SWAMP

Michael J. Duever[1]

ABSTRACT

The National Audubon Society conducted a comprehensive analysis of the Corkscrew Swamp Sanctuary ecosystem in southwest Florida during the 1970s. This work has provided a basis for the long-term management of the sanctuary in the face of an increasingly complex array of development pressures. The Corkscrew project gathered data on climate, hydrology, geology, soils, topography, vegetation, litter, nutrients, fire history, and human influences. From these data we have developed a detailed understanding, not only of the swamp's ecological characteristics, but also about how these characteristics interact and change in response to both natural and anthropogenic processes and events.

INTRODUCTION

The focus of this volume is the value of interdisciplinary research in dealing with the management needs of wetlands. Within this context, I would like to describe how we utilized a variety of scientific disciplines to develop a detailed understanding of the Corkscrew Swamp Sanctuary ecosystem. Much of the data generated by the individual disciplines has already been published elsewhere. In this paper I will provide only an introduction to the Corkscrew Swamp ecosystem and the various types of data we collected. The bulk of the paper will concentrate on discussing these data as they relate to the direct and indirect interactions of system components, with the objective of illustrating the importance of the interdisciplinary aspects of the study.

SITE DESCRIPTION

Corkscrew Swamp Sanctuary is located in southwest Florida along the northern edge of the Big Cypress Swamp. The National Audubon Society has been protecting Corkscrew Swamp since the 1920s. This was initially accomplished by employing game wardens, then by purchasing land, and most recently by actively managing it in the face of an increasingly complex mix of threats from agricultural and residential development on surrounding lands. Of particular concern is widespread drainage, which is gradually altering the hydrologic regime of much of the area (Duever et al., 1986). In addition to direct impacts on the ecosystem, the resulting drier conditions have altered the fire regime that existed prior to the recent intensive development of South Florida and caused more destructive wildfires (Wade et al., 1980). The spread of exotic plants and animals represents a third major threat to the integrity of the Corkscrew Swamp ecosystem.

---
[1]National Audubon Society, Rt. 6, Box 1877, Naples, Florida 33964

Most of Corkscrew Swamp is still in pristine condition, largely because of its position near the top of its watershed, which has minimized the impacts of man's drainage activities in the area. However, portions of the sanctuary have been impacted to varying degrees by logging, farming, off-road vehicle use, oil development, or grazing.

Climate

South Florida has a subtropical climate with hot wet summers, mild dry winters, and a spring drought. The frequent heavy summer rains are derived primarily from thunderstorms and are quite localized (Duever et al., 1986). Occasional tropical storms can contribute significant amounts of precipitation during late summer and early fall. Winter precipitation results from frontal systems and is much more evenly distributed over the region. Both hurricanes and occasional freezes can profoundly affect ecological patterns and processes.

Hydrology

The distinct wet-dry seasonal pattern of rainfall in South Florida controls the timing and degree of fluctuation in the water table (Parker, 1984). When the water table begins to rise in early June, the major cypress strands and deeper marshes act as broad shallow flowways, but as water levels continue to rise, these areas overflow into adjacent marshes and pinelands. During the summer rainy season, water moves in a generally southward direction as a thin, slowly-flowing sheet that covers most of the undeveloped interior of South Florida. Sometime in early fall the rains virtually cease, and water levels start to recede again into the major flowways.

Normal rainfall patterns result in a 0.5-1.5 m annual fluctuation in the water table at Corkscrew Swamp. However, water levels may be relatively stable in a year with abundant dry season precipitation or may drop as much as 2 m in a severe drought year (Duever et al., 1984). During extreme droughts, there is no natural surface water anywhere in Corkscrew Swamp.

Substrates

Bedrock is near the ground surface in only a few localized sites at Corkscrew Swamp, but nowhere have we found it more than 6 m below the surface (Duever et al., 1976). Surface soils are still simple geological and biological products that have not had sufficient time and appropriate environmental conditions for evolution into true soils, i.e. layers containing various mixtures of mineral and organic materials in characteristic profiles (Duever et al., 1986). The three major types of materials that make up the substrates at Corkscrew Swamp are sand, marl, and peat.

Sandy soils are found generally at higher elevations. They are derived primarily from old shoreline deposits and from rock weathering (Duever et al., 1986). Marls (calcitic clays) are found at intermediate to higher elevations, in areas dominated by short, sparse marsh vegetation and where limestone bedrock is near the ground surface. They are produced by physicochemical and biochemical processes of periphytic algae growing as mats on the bottom or as sheaths covering vegetation in freshwater wetlands (Gleason and Spackman, 1974). Peat formation is restricted to topographic depressions where litter decomposition is slowed by almost year-round inundation (Davis, 1946). In addition, sandy loams are found in some hardwood hammocks, and almost pure beds of marine shells can be found shallowly underlying some areas.

Vegetation

As with the entire Big Cypress Swamp, Corkscrew Swamp Sanctuary is dominated by extensive forests of cypress, Taxodium distichum, and a variety of emergent herbaceous marsh types (Duever et al., 1986). Uplands dominated by either slash pine, Pinus elliottii var. densa, or various combinations of temperate and tropical hardwoods (locally known as hammocks) occur as fringes around wetlands or as islands within them. A few small open water ponds (sloughs) are found within the cypress forest or deeper marsh areas.

Fire is one of the major factors that determines the character and distribution of South Florida plant communities (Wade et al., 1980). Except for the hardwood forests, all exist because of periodic fires, which keep them in a particular state of successional development. Even the character of hardwood forests is controlled by fire. They are typically young forests, which must become reestablished by root sprouting or the reintroduction of seeds, following occasional devastating fires during severe droughts.

Animals

The annual fluctuation in water levels above and below the ground surface over much of South Florida results in dramatic fluctuations in animal populations during the year (Carlson and Duever, 1977, Robertson and Kushlan, 1984). Animal adaptations to these varying conditions include rapid reproduction and growth, and either local or regional migration. By these means they are able to take advantage of resources, such as space and food, which are often only temporarily available. They can also avoid unsuitable environments, such as flooded or dry sites depending on whether they are upland or aquatic organisms, respectively.

METHODS

The Ecosystem Model

We began our research by developing a descriptive model of the Corkscrew Swamp ecosystem, including its major components, flows, and forcing functions (Figure 1). Major components were water, substrates (litter, peat, and soil), vegetation (overstory which included woody plants over 1 m tall, understory which included all other vegetation), and animals. Major flows of interest among the components, and into and out of the ecosystem, were water and nutrients. Major forcing functions influencing the state of components within the system were surface and ground-water flows, fire, day-to-day meteorological conditions, such as precipitation and temperature, and periodic catastrophic meteorological events, such as freezes and hurricanes. On the basis of this model and the resources available to us, we decided to initially focus our studies on the set of environmental parameters and processes shown in Table 1. We felt it was important to obtain at least limited information on all of these parameters to understand how each influences other portions of the system and how they, in turn, can be influenced by actual and potential management practices at this particular site. Most importantly, the breadth of this approach decreased the possibility of critical pieces of the puzzle "falling through the cracks" because we hadn't looked at a topic, even superficially.

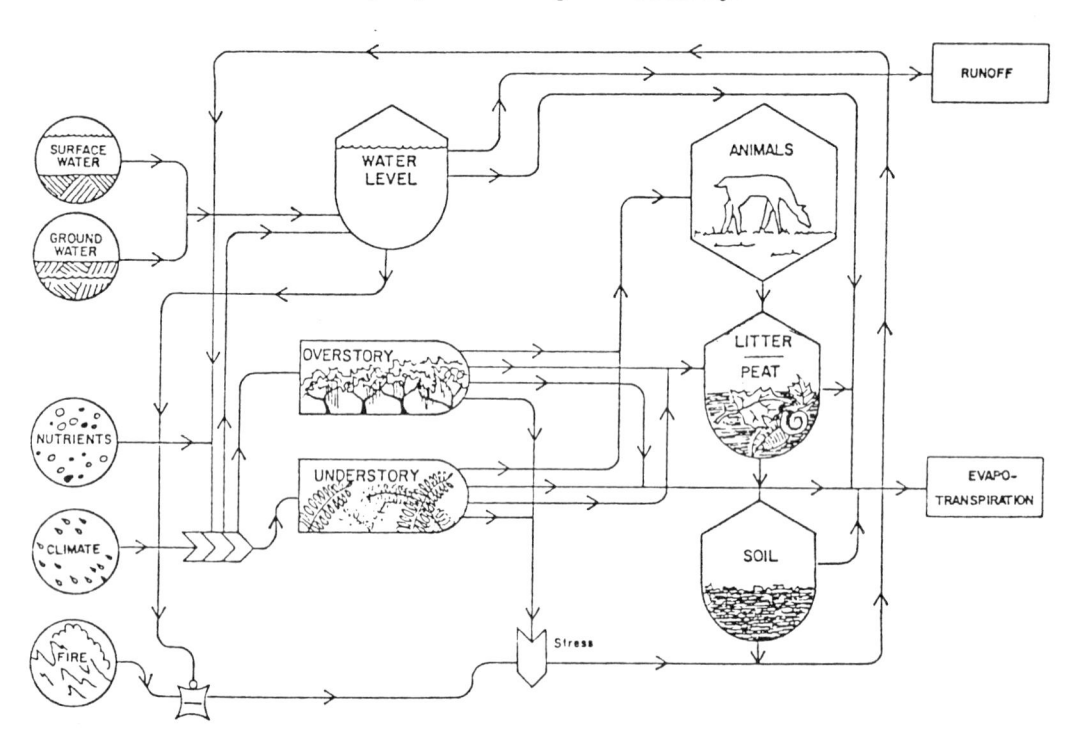

Figure 1. The descriptive model of the Corkscrew Swamp ecosystem that guided our initial research efforts. The components of the system include water, understory and overstory vegetation, litter and peat, soils, and animals. The forcing functions include surface- and ground-water inflows, nutrient inflows, climatic conditions, and fire. The flows are the arrows connecting the components and forcing functions, as well as the outputs from the system.

As new data became available, we periodically revised our model, primarily by expanding the details of what turned out to be particularly important components and/or flows. We then made

Table 1. Environmental parameters and processes studied in our ecosystem analysis of Corkscrew Swamp. One major aspect of our studies involved the evaluation of existing conditions, and another the evaluation of how past natural and anthropogenic events and processes have influenced the development of the existing plant communities.

EXISTING CONDITIONS

METEOROLOGY
    Precipitation
    Evaporation
    Temperature
        Maximum Daily
        Minimum Daily
    Relative Humidity

HYDROLOGY
    Water Levels
        Surface
        Ground
    Flows
    Evapotranspiration
    Water Quality

FIRE REGIME

SOILS
    Structure
    Chemistry

LITTER
    Litterfall
    Decomposition

VEGETATION
    Species Composition
    Biomass
    Productivity
    Chemistry

ANIMALS
    Fish
    Benthos

MAJOR CHANGES IN COMMUNITIES OVER TIME

SUCCESSION
    Vegetation Structure
    Annual Tree Growth Rings
    Peat
    Pollen

DISTURBANCE
    Fire
    Logging
    Logging and Fire
    Cattle Grazing
    Drainage
    Flooding

decisions about the adequacy of our or others' data for documenting the characteristics of these components and flows. For those portions of the model where the data were most inadequate, we attempted to devise approaches for obtaining the missing information. In this manner we increased or decreased our level of effort on certain aspects of the system and initiated several new studies.

Evapotranspiration from each of the major community types at Corkscrew Swamp is an example of a new study. It turned out to be the primary means by which water leaves the swamp, but there was only very limited information on the topic, particularly for wetlands. Another new study involved the relationship between growth rates of woody plants and the environmental conditions that influenced these rates. In the course of our research, we became aware of the potential for using annual ring width patterns of old-growth trees in the development of otherwise unavailable long term records of environmental parameters. We subsequently initiated a major study of these ring patterns in an attempt to develop proxy records for water levels, freezes, and hurricanes at Corkscrew Swamp over periods far beyond those available from historic records. We also expanded a 3 year study of the productivity of woody plants to a 10 year study of dbh (diameter at breast height) changes in about 6000 trees and shrubs to document growth rates, mortality, and recruitment of this relatively slow-growing component. These data are providing valuable information on the directions and rates of successional change in all of the major communities at Corkscrew Swamp.

In some cases, although the need was recognized, we have not been able to generate the desired data. One good example of this type of problem is information on roots and their role in the system. Roots are difficult to study under the best of conditions, but where the sites are under water for six or more months of the year, it is particularly difficult. As a result, we were not able to mobilize the resources or develop the techniques required to more than superficially address this need. However, since this component is explicitly recognized in our current model, we are using the best available information on its role in the system, and are clearly aware of the need to improve this aspect of our data base.

An important aspect of the iterative process of model development, literature review and/or field research, model revision, etc. is that an up-to-date model explicitly summarizes the current state of our knowledge about an ecosystem. The availability of this tool allows a scientist, at any stage of a research program, to respond to requests from managers about what is known about the significance of potential outside impacts on the system, as well as about the net effect of any proposed management activities. Early in a research program, most of our responses would have to be very general, and in many cases we would have to say we know little or nothing about even important aspects of the system. However, this type of interaction between managers and researchers is exactly what will maximize the usefulness of the model and the associated research program over the long run. This is because the interaction will help produce a model that is more compatible with real world management needs and practices. As the research program continues, we will increasingly be able to make more accurate statements about the significance of various management activities. These statements will not only be about probable effects of the practices on the system, but also about the likelihood of their achieving the manager's own objectives for restoring, maintaining, and protecting the various portions of the ecosystem.

Climate

The meteorological parameters monitored in our study were of significance primarily to the construction of a water budget

(Duever et al., 1975). For this reason, precipitation and evaporation were of major importance. Since Corkscrew Swamp is near the top of its watershed, most of the water entering the system originates from precipitation. Evaporation, as a component of evapotranspiration, is one of the primary mechanisms by which water leaves the system due to slow surface sheet flows and relatively high year-round temperatures (Duever et al., 1986).

Low temperatures can influence plant community composition because many tropical species are sensitive to South Florida's relatively mild and infrequent freezes. Measurement of maximum and minimum temperatures near ground level in the various habitats was particularly relevant because of the moderation of temperature extremes in areas with a forest canopy or standing water.

Hydrology

At Corkscrew, we monitored 31 wells for three years to characterize water-level fluctuations in eight major plant community types (Duever et al., 1975). When combined with topographic surveys of the transects along which the wells were distributed, these data allowed us to determine maximum and minimum water levels as well as hydroperiod for each community (Duever et al., 1978). Hydroperiod is the total number of days in an annual cycle during which there is water standing above the average ground surface elevation of a specific area. A three year evapotranspiration study was conducted at six sites representing the major community types on the sanctuary. We also estimated water flows into and out of the sanctuary as well as in the various plant communities.

One of the major products of our hydrologic studies at Corkscrew Swamp will be a water budget. Water flow data will allow us to estimate the movement of surface water through the system. Our precipitation and evapotranspiration data will allow us to account for atmospheric exchanges, and the water level data from monitoring wells will allow us to document changes in storage. The well data will also allow us to estimate net ground-water flows on the basis of water table recession rates.

We obtained surface- and ground-water samples from each of our wells for water quality analysis. When the data analyses for these samples are complete, they will contribute information necessary for the construction of nutrient budgets for the sanctuary's major plant communities as well as for the entire sanctuary.

Substrates

In studies at Corkscrew Swamp, we augered and described soil profiles in at least several examples of each major habitat type and in ecotones between them at locations throughout the sanctuary (Duever et al., 1976). Samples were also collected from representative profiles to quantify the structural and chemical characteristics of the major soil types (Coultas and Duever, 1984).

We determined depth to bedrock, where feasible with our equipment, and thickness of the organic soil layer at 30 m intervals along each of our transects. At several selected sites, cores of organic sediments were collected and subsampled for detailed structural, chemical, and pollen analyses (Stone and Gleason, 1976), as well as C-14 dating of a number of strata (Kropp, 1976; J. Taylor, 1980 unpublished report).

Vegetation

Along the same transects where we conducted our well monitoring and substrate studies, we established study plots to characterize each of the major types of plant communities on the sanctuary (Duever et al., 1975). Major community types included pine, hardwood hammock, mineral soil or organic soil emergent herbaceous marshes, mineral soil or organic soil shrub thickets, cypress swamp, and ponds. Within each of these types, there were variants that differed primarily as a function of substrate type, past disturbances, and/or recent successional processes. Within two replicate plots at each of our 31 sites, we recorded species composition and vegetation structure. Vegetation structure parameters included numbers and heights of individual woody trees and shrubs and diameters of each stem at 1.5 m (trees) or 30 cm above the ground (shrubs), average height of understory vegetation in 1 m^2 plots, aboveground and belowground dry weight biomass, and percent cover. We also estimated productivity and litter dynamics (litterfall, large woodfall, total litter on the ground, decomposition). At each site, we determined elemental composition of samples of leaves and wood from each woody species present, and of combined vegetation from the 1 m^2 understory plots. At selected sites, we conducted similar elemental analyses of total roots from each 15 or 30 cm stratum in root pits, of litterfall samples, of large woodfall samples, of ground surface litter samples, and of decomposition samples from each quarterly sampling period.

In addition to the intensive research done on our transect sites, several other vegetation studies contributed important supplemental information. In particular, analyses of cypress tree ring patterns allowed us to resolve a number of questions relevant to the structure of the swamp forest and how various environmental parameters influenced its productivity (Duever et al., 1976, 1978).

Animals

We collected data for one annual cycle on the numbers, biomass, and species composition of fish and benthic populations at three wetland sites along one of our intensive study transects (Carlson and Duever, 1979). While limited, these data did provide us with some information on animal population cycles in relation to the various environmental parameters as they exist at Corkscrew Swamp.

INTERACTIONS BETWEEN ECOSYSTEM COMPONENTS

Hydrology - Vegetation Relationships

Hydrology is the single most important factor controlling the distribution of plant communities at Corkscrew Swamp. The amplitude of annual water-level fluctuations above and below ground and the duration of water above ground interact with the adaptations for reproduction, germination, and growth of the various species to produce the swamp's communities. Maximum wet season water levels and, to a lesser extent, dry season minimum water levels are important, but hydroperiod proved to be a more significant factor (Duever et al., 1975, 1978). In general, it appears that sites with long hydroperiods cannot be colonized by species intolerant of extended inundation, while sites with short hydroperiods burn more frequently and severely, which periodically eliminates the more fire sensitive species.

The relatively greater significance of hydroperiod as compared to maximum or minimum water levels is based on a variety of observations from a number of sites at which we have worked. At Corkscrew Swamp, we found the major community types clustered closely on the basis of either maximum wet season water levels or hydroperiod (Duever et al., 1975). Although the same sites clustered together on the basis of minimum dry season water levels, each major community type exhibited a relatively greater spread of minimum dry season water levels. The increased scatter was a function of soil characteristics, which influenced water levels once they had receded below ground. This was particularly apparent with organic soils, which maintained higher water levels, and shell beds whose high porosity permitted a rapid outflow of water and decline of the water table.

Maximum wet season water levels that were 15-35 cm higher above a dike at Corkscrew Swamp have produced no dramatic differences in the plant communities above and below the dike after 10 years (Duever et al., 1975). Sufficient water seeps through the dike so that the hydroperiods are essentially the same on both sides. Also, the water table in the Okefenokee Swamp fluctuates much less than in Corkscrew Swamp, so that water levels there are neither as deep in the wet season nor as far below ground in the dry season. Yet hydroperiods are similar at both sites for comparable major habitat types (Duever, unpublished data).

The existence of anaerobic conditions in the soil is generally considered to be one of the most important factors controlling the distribution and kinds of wetland communities. The mere presence of water above the ground surface significantly reduces the movement of air into the soil. The longer this exchange is reduced, the more severe is the depletion of oxygen and the development of a reducing environment. This would suggest that conditions conducive to the development of different kinds of wetland vegetation are more a function of duration of inundation than depth of inundation.

One must be aware that the relative importance of different hydrologic parameters is a function of the limits within which each system operates. When these limits are exceeded, either naturally or through man's activities, the relative importance of parameters can shift. In addition, while certain parameters may be less important in certain situations, they can still significantly affect the system, such as influencing species composition or productivity, without necessarily altering the basic characteristics one uses to define the major community types. In South Florida or other areas where wetlands occur on broad expanses of very flat terrain, water levels are not likely to be that different from one place to another, but hydroperiods can vary tremendously and are more likely to influence the distribution of major community types. For comparable reasons, the distribution of wetland communities in areas with significant relief should more strongly exhibit the influence of maximum or minimum water levels, which are more variable than hydroperiod, due to the deeper and more ephemeral nature of the flooding.

Some Corkscrew Swamp plant communities are found over a broad range of average hydroperiods, while the ranges of other communities are relatively narrow (Figure 2). Upland pine forest and hardwood forest communities at Corkscrew Swamp typically have standing water for less than 2 months annually. Aquatic habitats, those lacking emergent vegetation, usually have standing water for more than 10 months. The transition between upland and marsh communities exists over a hydroperiod range of approximately 2-7 months, with the tallest and densest marsh vegetation found on sites inundated for 7-9 months. Although cypress were most often found on sites with hydroperiods of 8-10 months, those with the largest and oldest trees had hydroperiods of 9-10 months.

We were fortunate that the watershed upstream and downstream of the sanctuary was still intact in 1959 when National Audubon Society staff began to collect precipitation and water-level data at Corkscrew Swamp. Based on these data, we were able to document that a road and canal system, which came within 4 km of our downstream boundary by the mid-1960s, had not significantly altered Corkscrew's hydrologic regime (Duever et al., 1975). Thus, relationships shown in Figure 2, which were developed in the mid-1970s, have proved to be representative of a naturally-functioning South Florida wetland ecosystem.

Soil Influences on Hydrology

Information on soils at Corkscrew Swamp helped us interpret some of the variability we encountered in dry season water levels and ground-water flows (Duever et al., 1976). We found that water-level declines were more rapid on sites with mineral soils and slowest on sites with deep peat (Figure 3). The high capillarity of organic soils maintains the water table closer to the ground surface than do coarser soil types. Strata composed of almost pure marine shell are erratically distributed, but where they are extensive, their porous structure can result in rapid

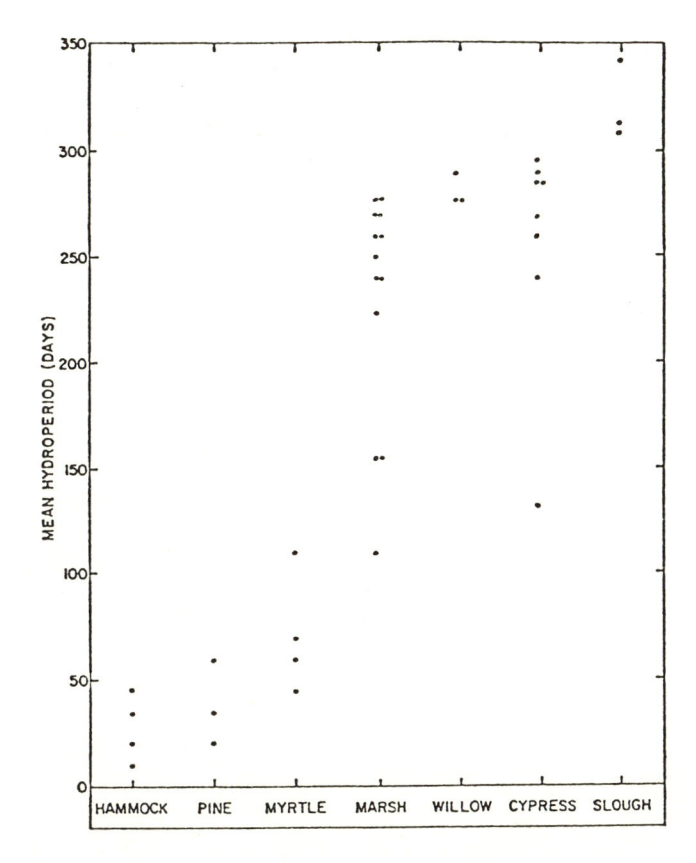

Figure 2. Hydroperiods for 38 sites representing all of the major types of plant communities at Corkscrew Swamp Sanctuary (Duever et al.,1978). The data are averages for 14 years of record.

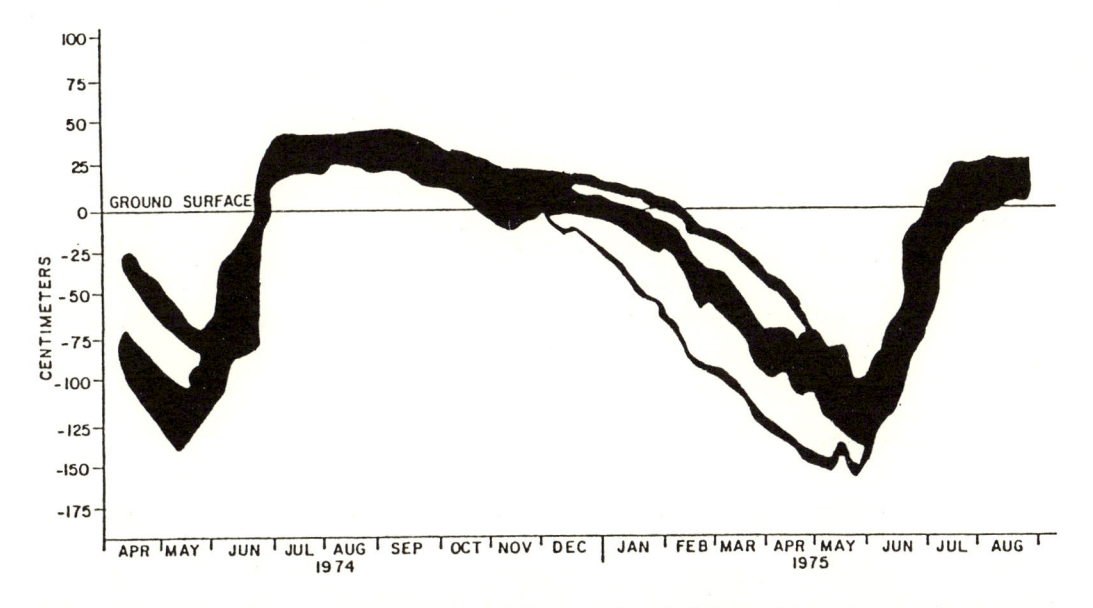

Figure 3. Water table elevations at eight well sites in marsh habitats at Corkscrew Swamp (Duever et al.,1975). The divergence in water levels between December 1974 and May 1975 reflects measurements from sites with more than 1 m of peat (upper band), 0.3-1 m peat (middle band), and less than 0.3 m peat (lower band).

localized declines in ground-water levels during the dry season
(Figure 4). Although layers of marl frequently underlie wetlands
in South Florida, they are rarely sufficiently continuous for
associated water tables to be considered perched, as in commonly
reported for wetlands.

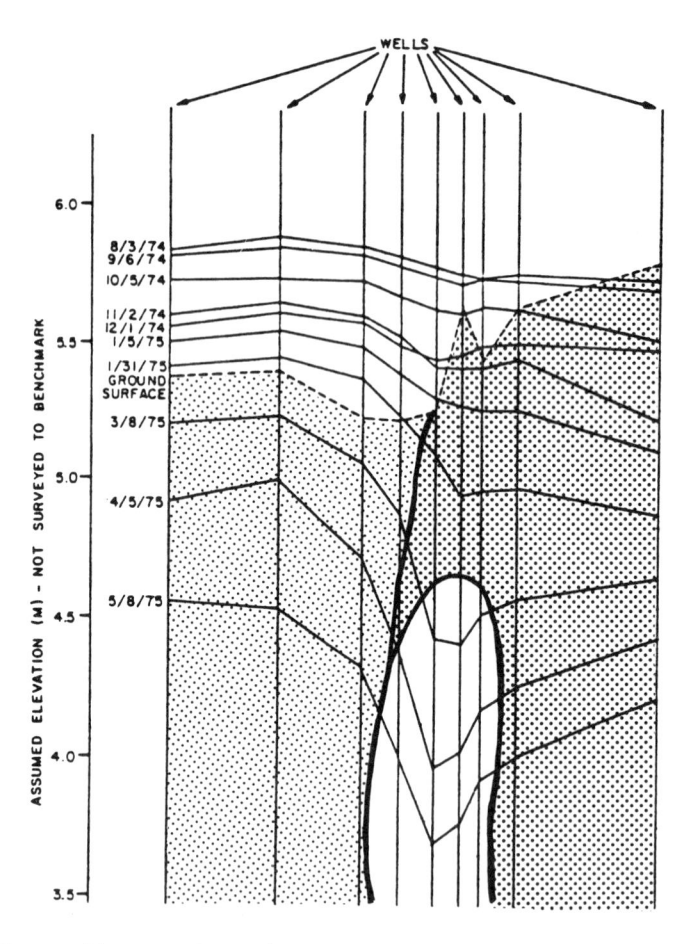

Figure 4. The pattern of 1974-1975 water-level decline along the
Central Marsh Transect at Corkscrew Swamp showing its relationship
to substrate characteristics (modified from Duever et al., 1975).
The different substrate types are indicated as follows: peat -
light shading (left); sand - dark shading (right); shell -
unshaded (middle bottom).

Soil - Vegetation Relationships

Soil characteristics at Corkscrew Swamp were found to
influence the composition and productivity of plant communities
rather than their basic type (e.g., pineland, hardwood hammock,
shrub, marsh, cypress swamp, or pond). There were three major
types of hammocks on the sanctuary. Swamp bay (Persea borbonia)
and red maple (Acer rubrum) hammocks were generally found on more
organic soils, oaks (Quercus laurifolia, Q. nigra, Q. virginiana)
dominated hammocks on deep sandy soils, and tropical hardwood
hammocks occurred on soils shallowly underlain by bedrock. In the

wetter marsh and cypress habitats, the deeper the soil and the more organic matter it contained, the denser and taller the vegetation was. Organic marsh soils also tended to support a community more dominated by sawgrass (Cladium jamaicense), arrowhead (Sagittaria sp.), and pickerelweed (Pontederia cordata), while sandy soils supported a more mixed community of grasses, sedges, and forbs. In somewhat later successional stages, these marsh sites supported willow (Salix caroliniana) and wax myrtle (Myrica cerifera) thickets, respectively.

Cypress Forest Gradients

We had two hypotheses for why dense stands of relatively small cypress trees occur at the outer edges of cypress forests in South Florida, and why their size increases while their density decreases toward the forest interior (Figure 5). One was that size differences are controlled primarily by site "quality", and the other was that the larger trees are merely older (Duever et al., 1984). The high correlation (r^2 = 0.915, n = 45) between size and age of trees along the intensively studied Central Marsh Transect through the eastern half of the cypress forest at Corkscrew Swamp indicated that age was a major factor (Figure 6). Examination of a number of environmental factors revealed no simple relationship between tree size (or age) and maximum wet-season water levels, minimum dry-season water levels, or ground surface topography. However, there was a high correlation (r^2 = 0.953, n = 18) between peat depth and dbh of the largest tree at each sampling point along the transect (Figure 7). A less-detailed examination of these same parameters in the western half of the same forest documented similar relationships there.

Figure 5. A profile from the edge (left) to the middle of the cypress forest along the Central Marsh Transect at Corkscrew Swamp showing the increasing height and size and the decreasing density of trees.

These correlations appear to be related to the annual duration of contact of the organic soil with the water table. In parts of the forest where the peat is deepest and is almost always in contact with the water table, the high capillary forces within the organic soils maintain relatively high moisture levels in the soils, litter, and vegetation, and a generally more moist microclimate within the forest. Particularly during the extended South Florida winter-spring dry season, a more moist environment would tend to reduce the frequency and severity of fires. This would, in turn, permit trees to persist for longer periods and attain greater ages and larger sizes.

As one moves toward the edge of the forest, the peat becomes more shallow and loses contact with the water table more frequently and for longer periods (Figure 6). The resulting drier conditions are therefore likely to produce more frequent and severe fires toward the edges of the forest, and the trees will be correspondingly younger and smaller. Beyond the strand edge, where cypress can germinate and survive on the basis of hydrologic conditions, fires are so frequent that the young trees are killed before they become well enough established to survive them.

Over the years, strands probably expand and contract in response to the occurrence of fire, which is largely controlled by long-term wet and dry cycles. During extreme droughts, even major strands are occasionally devastated by fire, but the forests are reestablished during subsequent wetter periods. Evidence of occasional devastating fires is suggested by C-14 dates for a layer of peat overlying an ash layer at the bottom of one of the deeper ponds within the cypress forest. This peat layer began to accumulate approximately 540 years B. P. (Pete Stone, pers. comm.). On the basis of ring counts of tree cores, we determined a maximum age of approximately 500 years for cypress from this same forest at Corkscrew Swamp.

Low-intensity fires are more frequent. Even in the undisturbed old-growth forests in Corkscrew Swamp, we found ample evidence (in the form of charcoal) of fires throughout the 5000-year-old peat profile (Stone and Gleason, 1976), as well as on surface materials which indicated the occurrence of fires within the last 50 years. Light burning generally has little effect on cypress trees, but it reduces the dominance of hardwoods in these forests and may eventually eliminate them (Ewel and Mitsch, 1978).

Cypress Age and DBH Relationships

When we plotted dbh against age for all of the available data on cypress at Corkscrew Swamp Sanctuary (Figure 8), we found there was a fairly consistent relationship (Coultas and Duever, 1984). Deviations from the relationship can be explained by differences in site conditions and histories of individual trees.

Trees growing in marl soils or in areas where rock is near the ground surface were generally small for their ages. This is

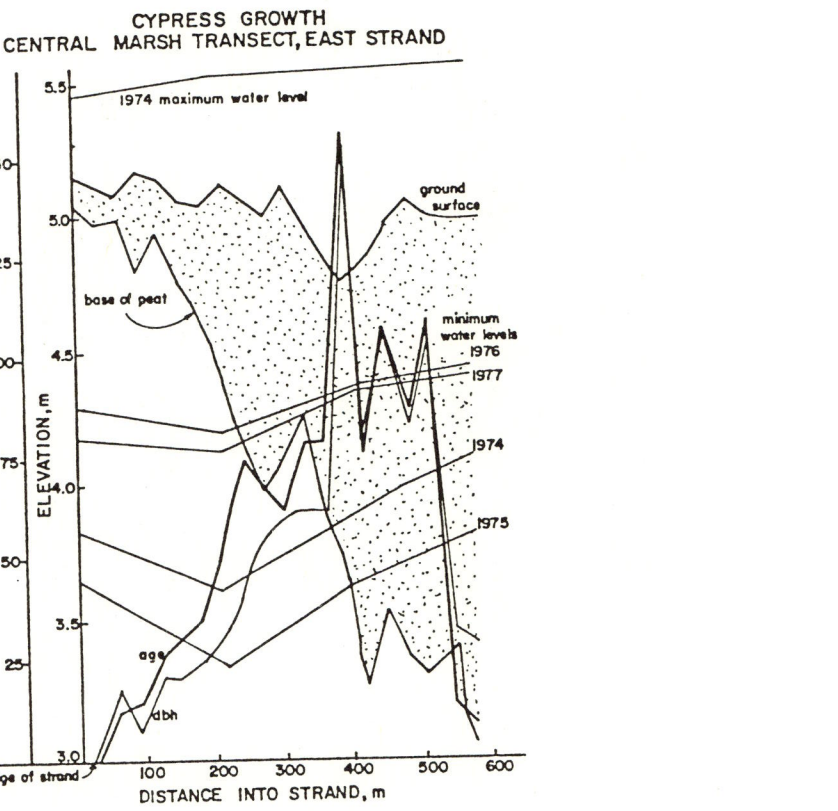

Figure 6. A profile through the same portion of the cypress forest illustrated in Figure 4 showing the elevation of the ground surface and base of the peat mass, 1974 maximum wet season water level, 1974-1977 minimum dry season water levels, and the average age and dbh of up to four cypress trees at 30 m intervals along the Central Marsh Transect (modified from Duever et al., 1984).

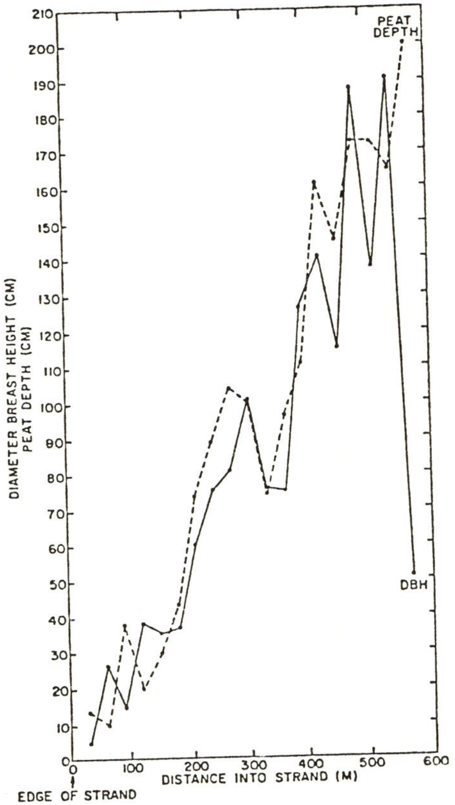

Figure 7. Peat depth and maximum cypress dbh at 30 m intervals along the same profile of the Central Marsh Transect shown in Figures 4 and 5 (Duever et al., 1986).

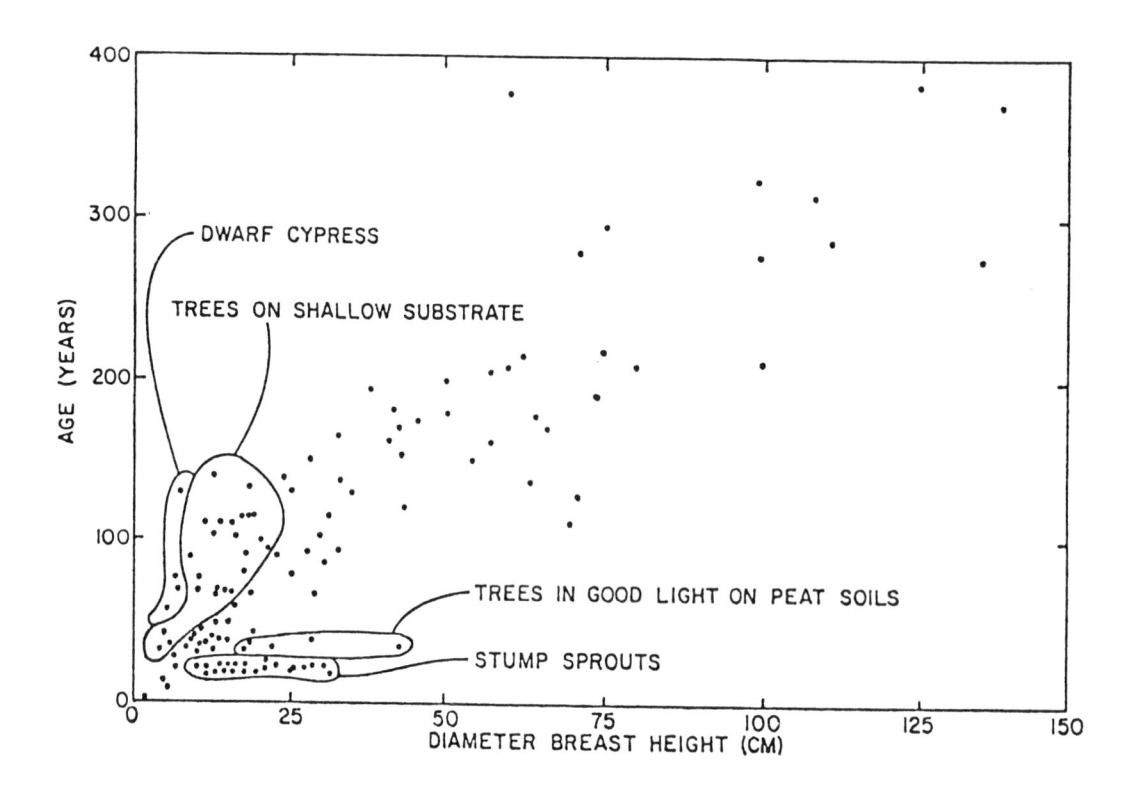

Figure 8. Correlation between cypress age and dbh for individual trees sampled at Corkscrew Swamp Sanctuary (Coultas and Duever, 1984).

most dramatic in the bonsai-like dwarf or scrub cypress with their frequently-enlarged, buttressed bases and small boles and crowns. Sites occupied by this cypress growth form are more shallowly-inundated than are sites occupied by cypress with more typical growth forms. Normally, this hydrologic regime should lead to relatively frequent fires and the elimination of these trees. However, the soil also limits the growth of the marsh vegetation that dominates these sites, and results in a slower buildup of fuels. This should produce fewer and less-severe fires, which in turn would allow a higher survival of cypress in what is otherwise an atypical environment, on the basis of its hydrologic characteristics.

Trees that were large for their age were from two distinct groups. One contained cypress that had resprouted from stumps in several logged areas. Not only did these trees have the advantage of the extensive root system of the original logged trees, but they had full access to sunlight in the cutover area. The second group contained isolated trees along the edges of the undisturbed swamp forest and in canopy openings associated with larger ponds within the forest. In these areas, increased light availability or reduced competition could account for their relatively rapid growth.

Effects of Altered Hydroperiod on Cypress Growth

As part of our efforts to confirm that cypress produces annual growth rings, we counted tree-rings on cores from stump sprouts of trees logged in the early 1950s. The trees were located immediately above and below an old elevated earthen logging road. The road had been rebuilt in 1966 to impound water within the sanctuary and protect it from what was then perceived as excessive water loss due to artificial drainage downstream. In the course of analyzing these samples, we noticed a number of interesting patterns in the tree rings (Duever and McCollom, in press).

In addition to the expected decline in growth rates with age, there were several abrupt growth changes that correlated with hydrologic conditions (Figure 9). Immediately after completion of the dike in 1966, there was a striking decrease in cypress growth rates. This was followed by a return to normal growth in 1968, and another striking decrease in 1969, which was maintained through 1975, the year prior to our collecting the cores. Sanctuary hydrologic records revealed that abnormally high water levels and long hydroperiods prevailed around the dike from 1967 through 1973. The only exception was 1968 when the dike was breached in several places by floodwaters and then repaired the following spring. A further indication of the relationship between extended inundation and reduced growth rates was that 1960 was the only year prior to construction of the dike when this site was naturally inundated year-round and when cypress growth was also extremely poor. Thus, it appears that even one year of continuous inundation was sufficient to drastically slow cypress growth.

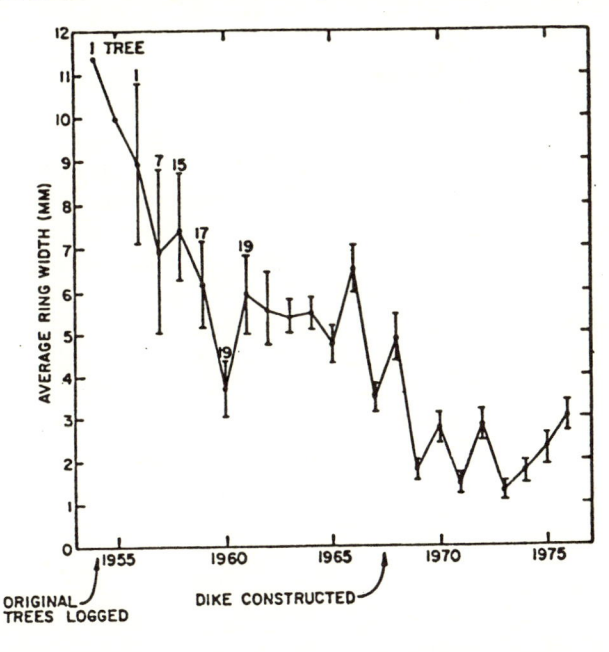

Figure 9. Mean ($\pm$ 1 S. E.) annual ring widths for cypress stump sprouts along the South Dike Transact at Corkscrew Swamp Sanctuary (Duever and McCollom, 1987). Except where data were available for all 20 trees, sample size is shown above each S. E. bar.

Although the trees immediately resumed rapid growth following a single year's flooding in 1960 and 1967, recovery after a more extended period of inundation apparently takes much longer. In 1974 and 1975, droughts permitted water levels to recede below ground each spring, but cypress growth rates remained below normal.

We sampled equal numbers of trees above and below the dike, primarily with the intention of generating replicate samples. However, during the data analyses, it became apparent that the patterns described above were applicable to both data sets. Since we had documented substantial differences in wet-season water levels above and below the dike (Duever et al., 1975), the only hydrologic parameter that was likely to be affecting both sides of the dike equally was hydroperiod. These values proved to be quite similar for both sides of the dike because of seepage through the porous substrate underlying it. Thus, the annual period of inundation appeared to be more significant in reducing cypress growth rates than was depth of inundation.

Successional Patterns and Rates

While the scheme depicted in Figure 10 does not include all the information that would be useful in describing successional patterns and rates in Big Cypress Swamp plant communities, it does illustrate how they relate to the two dominant environmental controlling factors, fire and hydroperiod (Duever, 1984). It also provides insights into long-term community changes that might be expected if these factors are altered by man's activities (such as drainage or impoundment projects or changes in burning frequency).

Ecosystem Change over Time

The areal extent, composition, and even existence of wetlands are all affected by long-term climatic cycles. While climatic conditions return to a more normal range of variability following a major drought or flood event, long term cycles can produce gradual, but over time, major shifts in the normal year-to-year range of hydrologic conditions. Thus, as climatic cycles become wetter, wetlands will tend to cover more of the landscape. There will also be a greater diversity of wetland communities because of a greater range of hydroperiods on different topographic sites and the development of deeper organic soils that will create new edaphic sites. A major difficulty in managing wetlands is our inability to distinguish between shifts in hydrologic conditions that result from man's activities and those resulting from sporadic natural events or long-term shifts in climatic conditions.

One of our major objectives at Corkscrew Swamp was to develop as long a record of environmental conditions as possible. Since historic records only went back as far as the late 1950's, we were forced to examine other sources of information to try to understand the time periods over which the Corkscrew Swamp

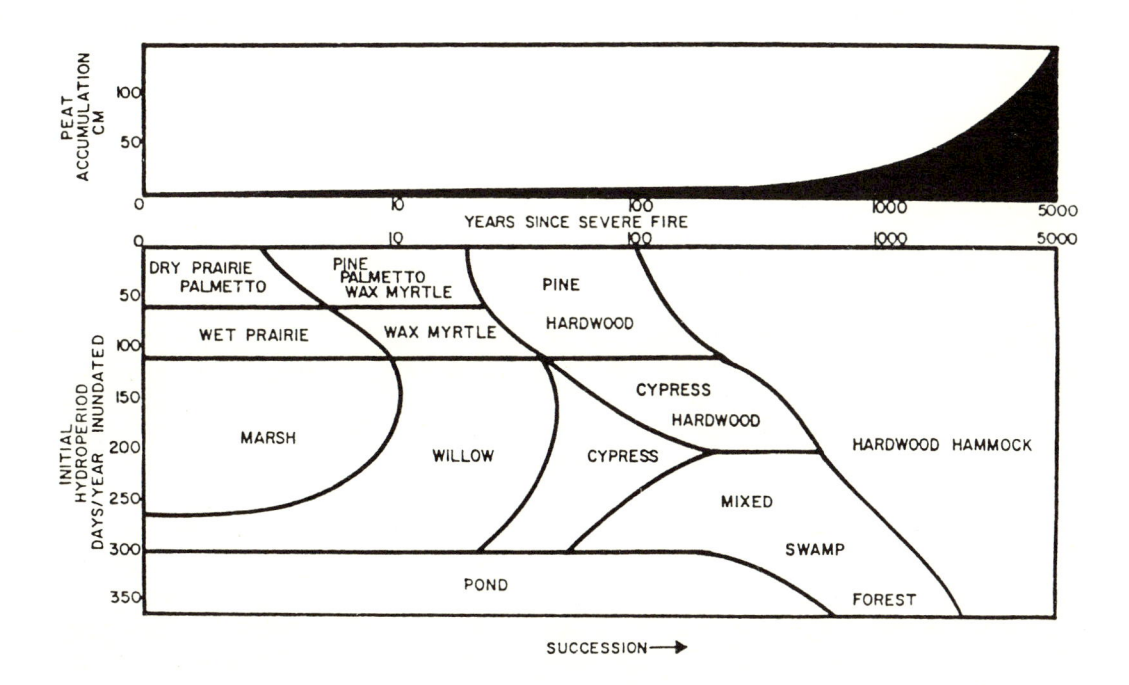

Figure 10. Successional patterns and rates of Big Cypress Swamp plant communities as a function of hydroperiod, peat accumulation, and time since severe fire (Duever, 1984).

ecosystem was changing, and the direction these changes were taking. To this end, we became involved in a variety of studies, including remote sensing, tree ring analyses, and studies of plant tissues and pollen contained in sediment cores.

Aerial photography was available for the whole sanctuary beginning in 1953 and at approximately 5-10 year intervals thereafter. Early 1940s photography was also available for portions of the area. These photos have proved valuable for documenting man's activities on and around the sanctuary. In an analysis of photography from 1953-1973, we did not detect any significant natural changes in the areal extent of plant communities (Gunderson, 1977).

In addition to the applications already discussed, tree rings can provide long term records for certain environmental parameters or events (Jacoby and Hornbeck, 1987). These records can be extended back even further than the age of living trees with the use of rings in wood preserved by either burial or submergence. The oldest living trees at Corkscrew Swamp are about 500 years old, but we have found stumps of dead trees overlain by approximately 2000 year old peat. However, while existing records provide data on conditions that occur on a daily, weekly, or monthly basis, trees produce only one ring each year. Thus, while they can provide relatively long-term records on the order of hundreds of years, their precision is only on the order of years.

Sediment cores involve yet other methodologies for learning about the past environmental history of a site. In this case, the time periods over which information can be generated are on the order of thousands of years, but their precision is on the order of hundreds of years. Cores taken at Corkscrew Swamp were from areas where fairly deep peat deposits are present. The materials we were working with were plant tissues and pollen preserved in the peat, and pollen in the marls that frequently underlie the peats. C-14 dating of samples from these deposits allowed us to determine not only the age of our current wetland system, but also the rate of vertical and horizontal development of the peat deposits that influence much of its character.

Approximately 10,000 years B. P., Corkscrew Swamp was a shallow marsh on a primarily marl and bedrock substrate (Kropp, 1976). By 5,000 years B. P., peats had begun to accumulate and have continued to steadily accumulate up to the present. As the peat mass grew vertically, it spread to an equivalent elevation over adjacent uplands (J. Taylor, 1980 unpublished report). Unfortunately, peats sampled to date have been too highly degraded to determine the type of wetlands that formed them, and the pollen samples still await analysis.

SUMMARY

Effective management of Corkscrew Swamp has required (1) understanding of the relative importance of a variety of major environmental parameters, (2) documenting processes that operate as thresholds or along gradients, (3) determining which parameters and processes are most sensitive to manipulation (i.e. disturbance or management), (4) evaluating the condition of each community relative to its original "undisturbed" condition, and (5) being able to differentiate natural long-term trends and cycles from man's influences. A particular advantage of the broad integrated approach used in the Corkscrew Swamp study was that we were able to identify the actual processes that control the major components of the sanctuary ecosystem. As a result, the sanctuary managers have been able to avoid wasting time and effort attempting to deal with what were merely symptoms or turned out to be insignificant or nonexistent problems.

ACKNOWLEDGMENTS

The author would like to thank Jean McCollom for reviewing the manuscript and assisting with preparation of the figures. Support for the original research was provided by the Rockefeller

Foundation through a subcontract with the University of Florida Center for Wetlands.

LITERATURE CITED

Carlson, J. E. and M. J. Duever, 1977. Seasonal Fish Population Fluctuations in South Florida Swamp. Proceedings of the Thirty-first Annual Conference Southeastern Association of Fish and Wildlife Agencies, pp. 603-611.

Coultas, C. L. and M. J. Duever, 1984. Soils of Cypress Swamps. In: Cypress Swamps, K. C. Ewel and H. T. Odum (Editors). University of Florida Press, Gainesville, pp. 51-59.

Davis, J. H., 1946. The Peat Deposits of Florida. Florida Geological Survey, Geological Bulletin 30, Tallahassee, 247 pp.

Duever, M. J., 1984. Environmental Factors Controlling Plant Communities of the Big Cypress Swamp. In: Environments of South Florida: Present and Past II, P. J. Gleason (Editor). Miami Geological Society, Coral Gables, Florida, pp. 127-137.

Duever, M. J., J. E. Carlson, J. F. Meeder, L. C. Duever, L. H. Gunderson, L. A. Riopelle, T. R. Alexander, R. F. Myers, and D. P. Spangler, 1986. The Big Cypress National Preserve. National Audubon Society, Research Report 8, New York, 444 pp.

Duever, M. J., J. E. Carlson, and L. A. Riopelle, 1975. Ecosystem Analyses at Corkscrew Swamp. In: Cypress Wetlands for Water Management, Recycling, and Conservation, H. T. Odum, K. C. Ewel, J. W. Ordway, and M. K. Johnston (Editors). Center for Wetlands, University of Florida, Gainesville, pp. 627-725.

Duever, M. J., J. E. Carlson, and L. A. Riopelle. 1984. Corkscrew Swamp: A virgin cypress strand. In: Cypress Swamps, K. C. Ewel and H. T. Odum (Editors). University of Florida Press, Gainesville, pp. 334-348.

Duever, M. J., J. E. Carlson, L. A. Riopelle, and L. C. Duever, 1978. Ecosystem Analyses at Corkscrew Swamp. In: Cypress Wetlands for Water Management, Recycling, and Conservation, H. T. Odum and K. C. Ewel (Editors). Center for Wetlands, University of Florida, Gainesville, pp. 534-570.

Duever, M. J., J. E. Carlson, L. A. Riopelle, L. H. Gunderson, and L. C. Duever, 1976. Ecosystem Analyses at Corkscrew Swamp. In: Cypress Wetlands for Water Management, Recycling, and Conservation, H. T. Odum, K. C. Ewel, J. W. Ordway, and M. K. Johnston (Editors). Center for Wetlands, University of Florida, Gainesville, pp. 707-737.

Duever, M. J. and J. M. McCollom, 1987. Cypress tree-ring analysis in relations to wetlands and hydrology. In: Proceedings of the International Symposium on Ecological Aspects of Tree-Ring Analysis, G. C. Jacoby and J. W. Hornbeck (Compilers). U.S.D.E., Washington, D.C., pp. 611-621.

Ewel, K. C. and W. J. Mitsch, 1978. The effects of fire on species composition in cypress dome ecosystems. Florida Scientist 41(1):25-31.

Gleason, P. J. and W. Spackman, Jr., 1974. Calcareous periphyton and water chemistry in the everglades. In: Environments of South Florida: Present and Past, P. J. Gleason (Editor). Miami Geological Society, Coral Gables, Florida, pp. 146-181.

Gunderson, L. H., 1977. Regeneration of cypress, *Taxodium distichum* and *Taxodium ascendens*, within logged and burned cypress strands at Corkscrew Swamp Sanctuary, Florida. Master's Thesis, University of Florida, Gainesville, 88 pp.

Jacoby, G. C. and J. W. Hornbeck (Compilers), 1987. Proceedings of the International Symposium on Ecological Aspects of Tree-Ring Analysis. U.S.D.E., Washington, D.C., 726 pp.

Kropp, W., 1976. Geochronology of Corkscrew Swamp Sanctuary. In: Cypress Wetlands for Water Management, Recycling, and Conservation, H. T. Odum, K. C. Ewel, J. W. Ordway, and M. K. Johnston (Editors). Center for Wetlands, University of Florida, Gainesville, pp. 772-785.

Parker, G. G., 1984. Hydrology of the pre-drainage system of the Everglades in southern Florida. In: Environments of South Florida: Present and Past II, P. J. Gleason (Editor). Miami Geological Society, Coral Gables, Florida, pp. 28-37.

Robertson, W. and J. Kushlan, 1984. The South Florida avifauna. In: Environments of South Florida: Present and Past II, P. J. Gleason (Editior). Miami Geological Society, Coral Gables, Florida, pp. 219-257.

Stone, P. and P. J. Gleason, 1976. The organic sediments of Corkscrew Swamp Sanctuary. In: Cypress Wetlands for Water Management, Recycling, and Conservation, H. T. Odum, K.C. Ewel, J. W. Ordway, and M. K. Johnston (Editors). Center for Wetlands, University of Florida, Gainesville, pp. 763-771.

Wade, D., J. Ewel, and R. Hofstetter, 1980. Fire in South Florida Ecosystems. United States Department of Agriculture, Forest Service General Technical Report SE-17, 125 pp.

TOLERANCE OF FIVE HARDWOOD SPECIES TO FLOODING REGIMES

L.H. Gunderson, J.R. Stenberg, A.K. Herndon[1]

ABSTRACT

Flood tolerance was assessed for seedlings of five hardwood species: Chrysobalanus icaco, Ficus aurea, Ilex cassine, Lysiloma latisiliquum, and Myrica cerifera. Treatments which consisted of watered-soil, flooded-soil, and flooded-seedlings were applied to potted seedlings for three months. Growth measurements (height, number of leaves, and biomass) and morphological adaptations were used to monitor performance under flood stress.

Growth and survivorship were highest among all five species in the watered-soil treatment. No growth was measured for any species in the flooded-seedling treatments. Flood tolerance was assessed by growth performance in the flooded-soil treatment and survival in the flooded-seedling treatment. L. latisiliquum was the least tolerant to flooding, as indicated by poor growth in the flooded-soil treatment and no survival in the flooded-seedling treatment. F. aurea, I. cassine, C. icaco, and M. cerifera all exhibited growth in the flooded-seedling treatment and survived three months of complete inundation.

Adventitious roots were produced by all species except L. latisiliquum. The number of roots produced by the Ilex and Myrica seedlings was related to the depth of flooding. The number of lenticels could not be related to flooding treatment.

INTRODUCTION

The hardwood wetland forests of the southern Everglades of Florida are locally referred to as tree islands. The name is descriptive of the clumps or patches of these swamp forests surrounded by marshes. Common hardwood species with both temperate and tropical affinities in these forests are redbay, Persea borbonia, sweet bay, Magnolia virginiana, wax myrtle, Myrica cerifera, dahoon holly, Ilex cassine, willow, Salix caroliniana, and cocoplum, Chrysobalanus icaco, (Davis, 1943; Loveless, 1959; Craighead, 1971; Alexander and Crook, 1975).

The distribution of southern Florida wetland hardwood species is primarily controlled by the hydrologic regime, although other factors such as fire history, edaphic conditions, and proximity to seed sources are also important. The hydrologic regime within undisturbed areas is a function of the topography and the annual rainfall pattern. Higher elevations are generally associated with outcroppings of surficial bedrock, lower sites with troughs in the bedrock. Local soil surface elevation may also be modified by organic soil accumulations. Most (80%) of the annual rainfall

---

[1] South Florida Research Center, Everglades National Park, P.O. Box 279, Homestead, Florida 33030

(mean = 140 cm) occurs during the summer months, with peaks in June and August. The resulting inundation pattern in these forested wetlands is characterized by mean inundation periods of 2 to 4 months, and water depths of 0 to 50 cm (McPherson, 1973).

Large-scale water management practices have altered the hydrologic patterns in most of the wetlands of southern Florida. The Central and Southern Florida Flood Control Project was initiated in 1949 and has created a series of levees, canals, pumps, and water control structures. The primary objectives of the project are flood control and maintenance of the water supply. Canals rapidly route water and are used to dissipate flood waters, whereas water is impounded in shallow reservoirs for use during the dry season. These practices alter natural seasonal variation in water levels and have affected vegetation throughout the Everglades region, with dramatic effects on the swamp forests or tree islands.

Drier conditions, due to either naturally occurring periods of low rainfall or diversion of water for certain management objectives, can cause an increase in the size of individual tree islands. Johnson (1958) documented the spread of "bushy" vegetation between 1940 and 1952 due to canal overdrainage. Kolipinski and Higer (1968) analyzed aerial photographs taken in 1940, 1952, and 1962 over Everglades National Park and measured an increase in tree island acreage over this time period. Correlated with the decrease in depth and duration of flooding is an increase in frequency and severity of fires. Fires can modify species composition by changing edaphic and subsequent hydrologic regimes by removing organic soil and lowering the soil surface elevation. Floods have been and are recurring phenomena that have influenced the species composition within these hardwood swamp forests. Management actions which aggravate these flooding conditions can influence tree island size and structure. Craighead (1971) found that abrupt and prolonged (several months) flooding due to regulatory water releases into Everglades National Park caused a die-off of hardwoods in tree islands. McPherson (1973) documented the mortality of hardwood species as a result of increased flooding (depth and duration) and subsequent decrease in areal extent of the island.

Community level responses to alterations in hydrologic regime have been documented, but very little information exists on the specific adaptations of southern Florida species to flooding. Quantitative data on flood tolerance of swamp forest species will be useful in evaluation of the effects of water management actions on these forests, especially which species may be most sensitive to flooding. We examined the responses of seedlings of five hardwood species which were subjected to a range of flooded conditions. We measured survival and growth of seedlings in each of two depths of flooding and one no-flood condition, in order to determine which species survived and grew under these hydrologic conditions. We also monitored each species for morphological adaptations to flooding. Adventitious root development and lenticel hypertrophy are two common flood adaptations of woody plants (Kozlowski, 1984).

METHODS

Five evergreen species (three swamp forest and two upland types) were selected for the study. The three swamp species used were: dahoon holly, _Ilex cassine_; wax myrtle, _Myrica cerifera_; and cocoplum, _Chrysobalanus icaco_. The upland species were: wild tamarind, _Lysiloma latisiliquum_, and strangler fig, _Ficus aurea_. Seedlings for the experiment were obtained from local nurseries and were germinated from native genetic stock. All seedlings were obtained during October 1983 and all species except _Myrica cerifera_ had been germinated during the preceding six months. Since seed from these different species ripen at different times through the spring and summer, the seedling age ranged from approximately two to six months. The _Myrica_ "seedlings" appeared to actually be cuttings, but we were unable to verify this with the nursery. All seedlings were less than 20 cm tall.

Eleven seedlings of each species were planted in a circular arrangement within a round, galvanized tub. Dimensions of the tub were 55 cm (bottom inside diameter) and 67 cm (top inside diameter) and 29 cm deep. The seedlings were planted in a 10 cm deep potting soil/ peat mixture during December 1983. The seedlings acclimated in drained tubs for approximately 1.5 months prior to the initiation of treatments. Each tub was used as a treatment unit with no replicates per treatment.

The experiment consisted of subjecting each species to three treatments of flooding for a three month period. The driest treatment maintained the soil moisture at field capacity (designated the watered-soil treatment). The intermediate treatment held water levels at the soil surface (flooded-soil treatment). The wettest treatment consisted of filling the tubs to capacity (19 cm standing water) and was called the flooded-seedling treatment. The water levels were maintained in the tubs by drilling holes at the appropriate elevation. Holes were made in the bottom of the driest treatment tub and at the soil surface of the intermediate treatment. Each tub was watered at least once every two days to maintain the desired water level. The water in the flooded tubs was drained and replaced every month in order to make measurements and harvest seedlings. The fifteen tubs were arranged in a 5 x 3 grouping. The species and treatments were given random assignments within the arrangement. The uncovered tubs were located in a fenced, paved area adjacent to the South Florida Research Center, Everglades National Park. The treatments were initiated on 15 February 1984 and continued through 18 May 1984.

Seedling growth was assessed by measuring three parameters: height, number of leaves, and plant mass. One seedling from each treatment tub was randomly selected and harvested for biomass determination on each of the following dates: 15 February, 14

March, 12 April 1984. Four seedlings were randomly selected and harvested from each tub on 18 May 1984. Harvested seedlings were separated into component groups of leaves, stems, coarse roots, and fine roots, then each group was oven-dried at 70°C for two days and weighed to the nearest 0.01 g. Heights of the remaining seedlings were measured (to the nearest 0.5 cm) and number of leaves/seedling was counted on measurement days.

We monitored morphological responses of the seedlings to flooding by counting the number of adventitious roots and lenticels on each seedling. The roots were counted on all remaining seedlings on each harvest date. Lenticels were counted using a Bausch and Lomb dissecting microscope (3x) on each harvested seedling along a 5 cm longitudinal strip (initiating at the top root and extending in an apical direction).

RESULTS

Seedling Survival

Seedling survival after three months was surprisingly high among species and treatments. All of the Ficus, Ilex, Myrica, and Chrysobalanus seedlings survived the watered-soil and flooded-soil treatments (Table 1). Seedling mortality, defined by the absence of leaves, occurred as a result of total immersion of each of these four species. Ilex seedlings had the highest survival (88%) in the flooded-seedling treatment, with fewer surviving Myrica (75%), Ficus (63%), and Chrysobalanus (50%)(Table 1). Some of the Myrica and Ficus seedling survived the flooded-seedling treatment by growing apical meristems to approximately 1 cm above the water surface. Lysiloma seedlings were most sensitive to all flooding treatments; 75% survived the watered-soil and flooded-soil treatments and none survived the flooded-seedling treatment.

Table 1. Seedling survival after three months of flooding treatments. Values are percent of remaining seedlings with leaves.

SPECIES	Watered-Soil	Flooded-Soil	Flooded-Seedling
Ilex cassine	100	100	88
Myrica cerifera	100	100	75
Ficus aurea	100	100	63
Chrysobalanus icaco	100	100	50
Lysiloma latisiliquum	75	75	0

Growth Measurements

Seedling growth measurements indicated differential species response to the treatments. Total biomass growth (including fine and coarse roots, stems, and leaves), for a given species generally was highest in the watered-soil treatments and was negligible in the flooded-soil and flooded-seedling treatments. Measurements of biomass, leaf-count, and height did not necessarily show the same response for a given species to a flooding treatment.

Lysiloma seedlings exhibited little or no growth in any of the treatments (Figure 1). Mean total biomass from the May sample was not significantly different (ANOVA, P=0.05) from the February sample in any of the treatments. Flooded seedlings lost all of their leaves within three days and did not change in height. Although no direct measurement of death was made, these seedlings were considered to be dead. Number of leaves and seedling heights were not significantly different (ANOVA, P=0.05) among the treatments.

Mean biomass per Ficus seedling was greater in the May watered-soil samples than in the February samples (Figure 2), but high variance precluded a statistically significant difference between the two times. Seedling height data tracked the biomass data, with an insignificant increase in height in the watered-soil treatment and no change in height in the flooding treatments. Number of leaves per seedling was significantly higher in May than in February (ANOVA, P=0.05) in all of the treatments. Some of the seedlings in the flooded-seedling treatment grew above the water level, while some remained submerged. The emergent seedlings grew additional leaves, which resulted in a higher treatment mean that was not representative of the treatment. These data indicate that Ficus does not grow under flooded conditions.

Mean Chrysobalanus seedling biomass was not significantly different between February and May in any of the treatments (Figure 3). Number of leaves per seedling was significantly higher in the May than in the February sample (ANOVA, P=0.05) in both the watered-soil and flooded-soil treatments. Seedling height remained unchanged in all tubs from February through May. Ilex seedlings increased in height and number of leaves in the watered-soil and flooded-soil treatments, but high variance masked significant differences in treatment effects (Figure 4). Leaf number dropped significantly in the flooded-seedling treatment, while heights and mass remained constant. These results in the flooded-seedling treatment are explained if the bulk of the mass is attributed to stems and roots, not leaves. The data indicate that Ilex has the ability to not only survive complete submergence, but to grow (increase leaf number and height) in a flooded-root environment.

Myrica seedlings grew substantially in the watered-soil and flooded-soil treatments. Mean total biomass/seedling, number of leaves, and seedling height all significantly increased (ANOVA, P=0.05) from February through May (Figure 5). Mean total seedling biomass and seedling height stayed the same in the flooded-seedling tubs, while leaf number decreased as a result of this treatment (Figure 5). These results indicate that Myrica is the

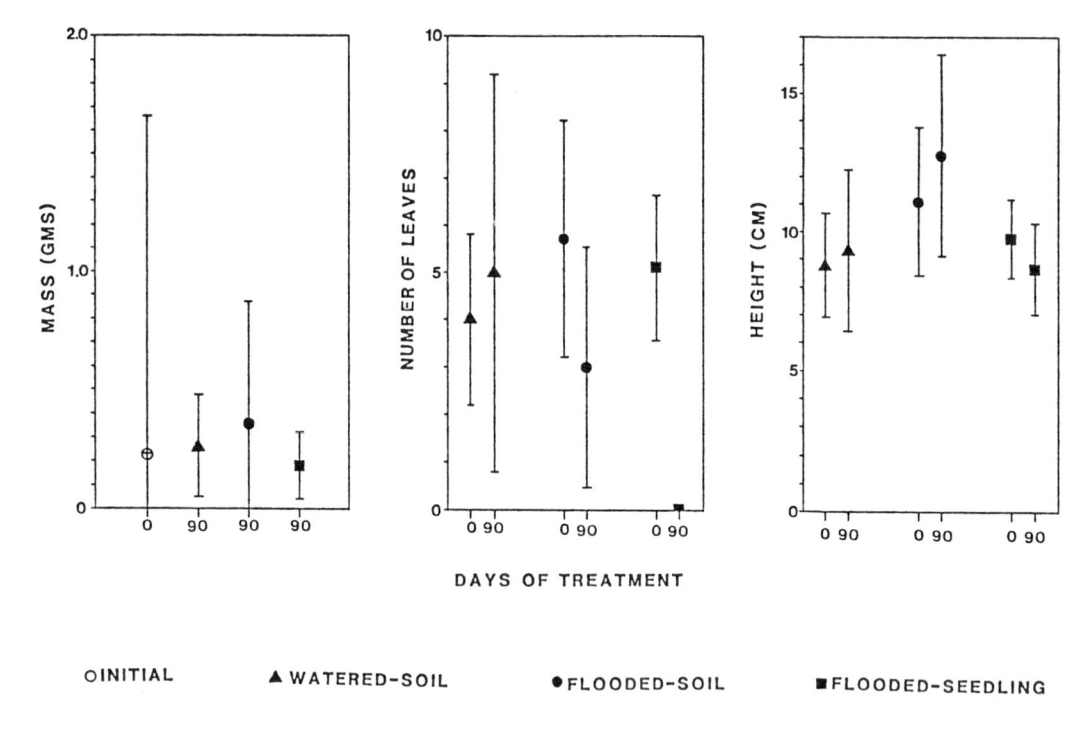

Figure 1. Total biomass, number of leaves, and height per seedling of <u>Lysiloma latisiliquum</u> at the beginning (n=3) and end (n=4) of 90 days of treatments. Values are means, ranges are 95% confidence intervals.

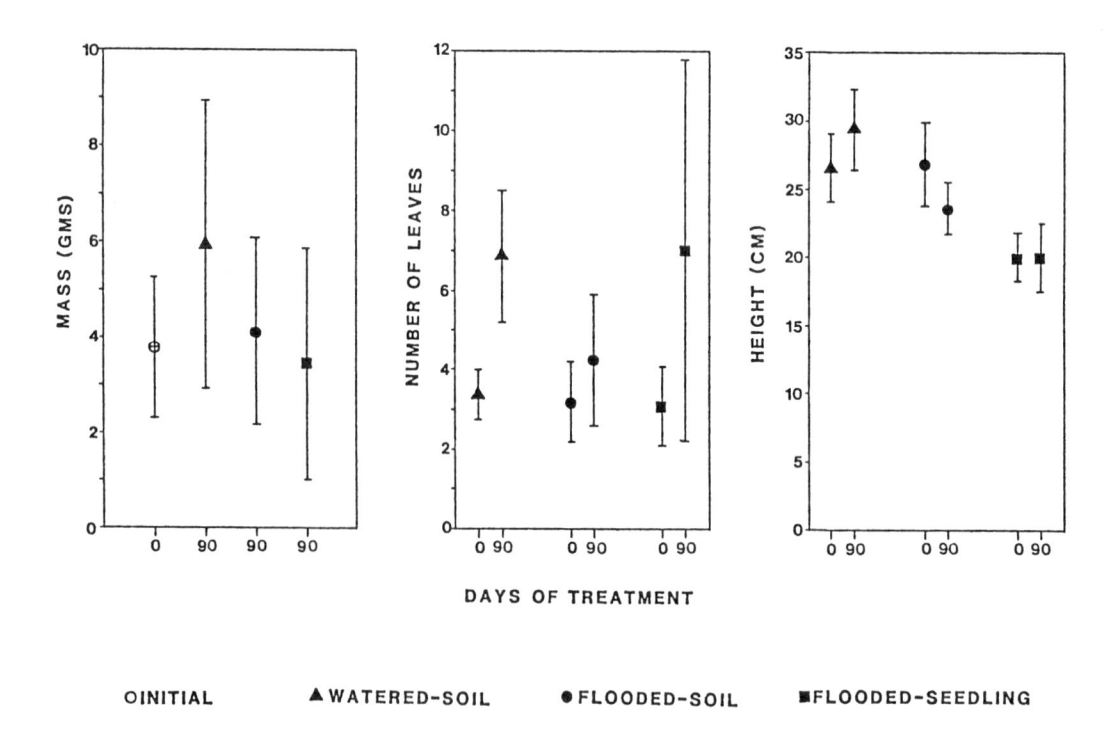

Figure 2. Total biomass, number of leaves, and height per seedling of <u>Ficus aurea</u> at the beginning (n=3) and end (n=4) of 90 days of treatments. Values are means, ranges are 95% confidence intervals.

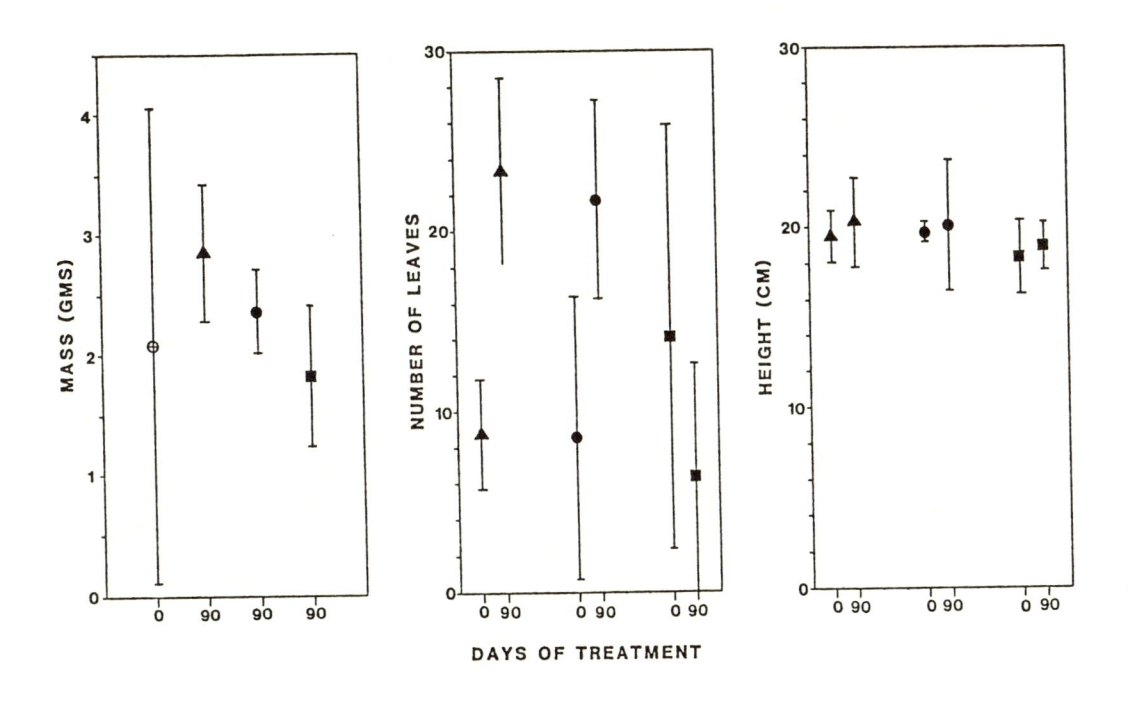

O INITIAL   ▲ WATERED-SOIL   ● FLOODED-SOIL   ■ FLOODED-SEEDLING

Figure 3. Total biomass, number of leaves and height per seedling of <u>Chrysobalanus icaco</u> at the beginning (n=3) and end (n=4) of 90 days of treatments. Values are means, ranges are 95% confidence intervals.

O INITIAL   ▲ WATERED-SOIL   ● FLOODED-SOIL   ■ FLOODED-SEEDLING

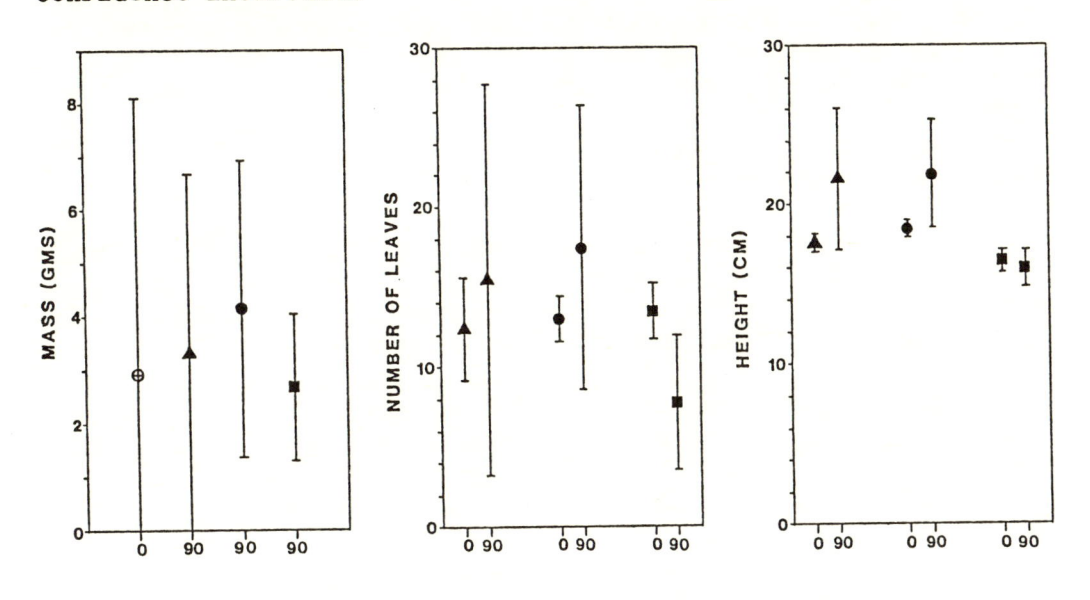

Figure 4. Total biomass, number of leaves and height per seedling of <u>Ilex cassine</u> at the beginning (n=3) and end (n=4) of 90 days of treatments. Values are means, ranges are 95% confidence intervals.

most flood-tolerant of the species tested by exhibiting significant growth (both increasing leaf number and height) in the flooded-soil treatment.

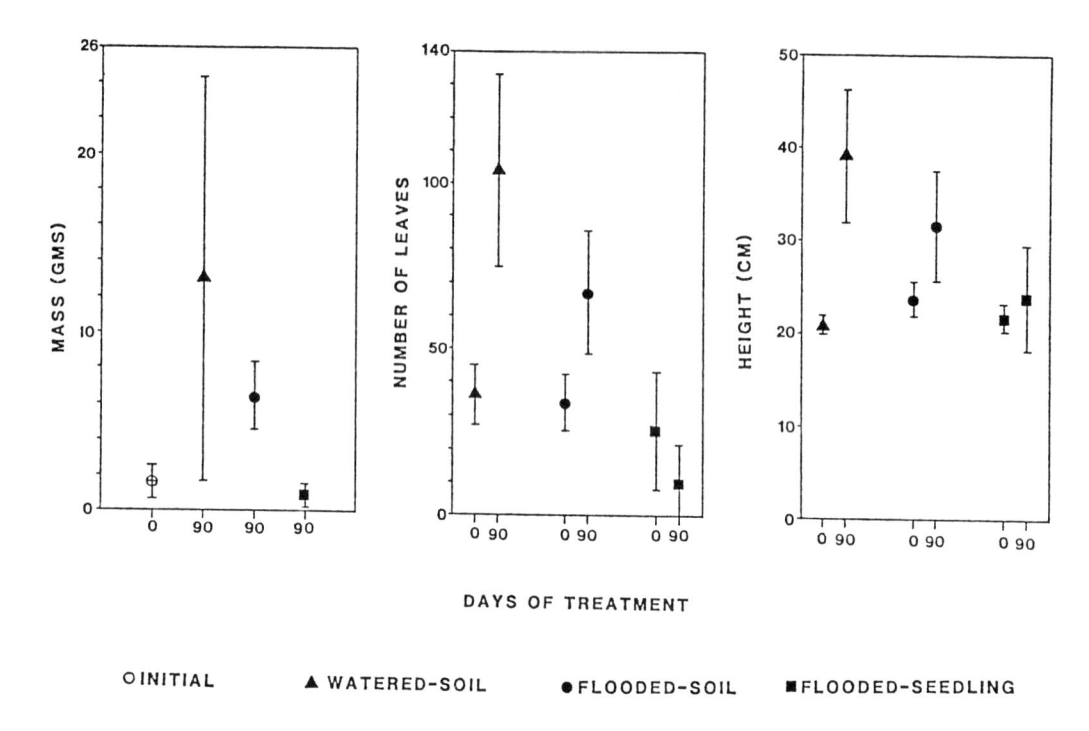

Figure 5. Total biomass, number of leaves and height per seedling of Myrica cerifera at the beginning (n=3) and end (n=4) of 90 days of treatments. Values are means, ranges are 95% confidence intervals.

Morphologic Responses

The number of adventitious roots varied among species and treatments. The Chrysobalanus and Lysiloma seedlings showed no adventitious root response to treatments, even though the Chrysobalanus had roots present in all treatments (Table 2). The Ficus seedlings generally had the same number of roots/seedling at the beginning and end of the experiment in the two flooded treatments, but showed a slight decrease in the watered-soil treatment. None of the Ilex or Myrica seedlings had adventitious roots in February, and none of the seedlings of either species in the watered-soil tubs had produced adventitious roots by May. However, by May, Ilex had only slightly more and Myrica had substantially more adventitious roots in the flooded-seedling tubs than in the flooded-soil treatments.

No clear trends emerged from the lenticel count data (Table 3). Ficus and Lysiloma had higher counts in all treatments from the May sample than the February sample. The number of lenticels that were counted on Ficus during May was lowest in the watered-soil and highest in the flooded-seedling tubs, while the opposite trend was observed in Lysiloma. The mean number of lenticels on the Ilex seedlings decreased in all of the treatments. Due to small sample sizes and high variability, none of the observed differences were statistically significant.

Table 2. Mean number (± one std. dev.) of adventitious roots per seedling during the harvest months of February and May.

SPECIES	MONTH	Watered-Soil	Flooded-Soil	Flooded-Seedling
Lysiloma latisiliquum	Feb.	0	0	0
	May	0	0	0
Ficus aurea	Feb.	2.3 ±1.5	3.6 ±1.8	2.8 ±1.8
	May	0.4 ±0.7	2.8 ±1.4	1.8 ±1.8
Chrysobalanus icaco	Feb.	1.4 ±1.4	0.9 ±1.4	0.8 ±1.0
	May	1.1 ±1.6	1.0 ±1.4	0
Ilex cassine	Feb.	0	0	0
	May	0	0.3 ±0.5	0.5 ±1.0
Myrica cerifera	Feb.	0.3 ±1.0	0	0
	May	0	1.3 ±2.0	3.8 ±7.0

Table 3. Mean number (± std. dev.) of lenticels per 5 cm of stem harvested during February and May.

SPECIES	MONTH	Watered-Soil	Flooded-Soil	Flooded-Seedling
Lysiloma latisiliquum	Feb.	--	15.0 ±8.0	--
	May	23.5 ±6.5	20.5 ±5.0	15.5 ±2.3
Ficus aurea	Feb.	--	6.4 ±2.4	--
	May	10.0 ±4.8	11.5 ±3.6	13.0 ±4.5
Chrysobalanus icaco	Feb.	--	16.0 ±6.0	--
	May	15.5 ±6.2	10.3 ±1.3	13.4 ±6.6
Ilex cassine	Feb.	--	25.0 ±14.5	--
	May	10.5 ±2.5	17.5 ± 6.3	13.5 ±6.2
Myrica cerifera	Feb.	--	11.2 ±2.8	--
	May	9.5 ±2.0	18.5 ±8.0	9.8 ±1.0

DISCUSSION

The genetic traits and age of individual plants can influence a species' ability to tolerate flooding (Gil, 1970). Site-related factors such as soil type, depth of flooding, duration of flooding, and water quality parameters can also determine responses of plants to flooding. Assuming that the age, genetic stock (within a species), soil, and water quality were consistent among the applied treatments, this experiment allowed a relative assessment of response to depth of flooding.

Seedling Survival

The net result of a species' morphological and metabolic adaptations to a flood event is the ability to survive. Briscoe (1957) found that severity of effects increases with duration and depth of flooding. He found that many species will survive soil flooding even though growth rates are decreased, and that total submersion of a species generally results in death. Putnam (1951) found that no species other than willow (Salix sp.) can survive submerged foilage more than a few days. Demaree (1932) and Gunderson (1977) found cypress (Taxodium sp.) seedlings died within 4 to 6 weeks of total immersion. Loucks and Keen (1973) did document survival of ash, Fraxinus caroliniana, and cypress after a 4-week period of total immersion. These reports indicate some woody species can survive at least one month of submergence, while longer periods generally result in death.

Our data indicated that survival of the tested species was surprisingly high for a 3-month period of submergence. Fifty percent or more of the individuals of four of the five tested species survived (Table 1). We attribute the survival to a combination of reasons: the ability of the species to survive inundation and the method of treatment. The treatment method consisted of draining and rapidly refilling the tanks. By draining the tanks, short periods of oxygenation probably occurred. Refilling the tanks with "new" water probably decreased the concentrations of phytotoxins that may have increased since the last refill. These results imply that the renewal or turnover rate of the water may influence the survival of these hardwood seedlings, possibly by decreasing the concentrations of toxic anaerobic by-products.

Growth Responses

Leaf chlorosis, epinasty, and abscission are all commonly observed responses to flooding. These effects are a result of loss of viable roots and resultant impairment of transpiration and translocation (Kozlowski, 1984). The number of leaves per seedling varied with the treatments and corroborated the other growth parameters, biomass and height, as a response to flooding. Leaf number always increased in the watered-soil treatment and generally decreased in the flooded-seedling treatment. All of the leaves abscised in the flood-intolerant L. latisiliquum and some abscission occurred in all of the other species in the flooded-seedling treatment.

Biomass accretion under flooded conditions is limited by the lack of adaptations of plant systems to operate under anaerobiosis

(Hook and Scholtens, 1978). Impaired root metabolism results in closure of stomata, thereby decreasing photosynthesis (Sena-Gomes and Kozlowski, 1980). Root metabolism can be impaired by any or all of the following reasons: lack of oxygen diffusion to parts of the root system, inability to utilize alternative pathways of respiration, and the accumulation of phytotoxins that are by-products of anaerobosis, (Armstrong, 1968; Kozlowski, 1984). By following the total biomass changes and the ratio of above to below ground plant mass, inferences can be made as to the flood tolerance of the studied species.

Since insignificant growth (as measured by both height and mass increase) occurred in the flooded-seedling treatments, we chose to evaluate flood tolerance based upon changes in height and mass after 90 days in the flooded-soil treatments. No statistically significant growth was measured in L. latisiliquum, F. aurea, nor C. icaco. I. cassine had a significant increase in height, indicating some tolerance of root flooding. The M. cerifera seedlings had statistically significant increases in mass and height, indicating the greatest tolerance.

The ratio of below to above ground biomass (root/shoot ratio) has been used to track seedling response to fluctuating water levels. Schwintzer and Lancelle (1983) found the root to shoot quotient of Myrica gale decreased with increasing water depth. This response was due to an ability to sustain less root biomass under wetter conditions with less oxygen. Kummerow (1980), however, found the root to shoot ratio to be variable among species and not necessarily related to water stress. Our data support Kummerow's contention, as indicated in Figure 6. No consistent trends were evident in the root/shoot relationship among our tested species that could be attributed to the flooding treatments. Ficus seedlings had the largest ratios, which is consistent with the observed number of adventitious roots produced. This species appears to allocate biomass to roots more than above-ground portions. Ilex seedlings in the flooded-seedling treatment (Figure 6) had a high root/shoot ratio, which we attribute to die-back of the above-ground mass, as well as some root production. Even though Myrica produced numerous adventitious roots, the root/shoot ratio did not change.

Morphological Responses

Adventitious roots (flood roots) are a common response of inundated woody species under anaerobiosis (Hook and Scholtens 1978) and compensate for the absorption lost by the original root system (Kozlowski 1984). The presence or absence of flood roots is a function of both a genetic tendency to produce roots and environmental conditions that trigger the response. Four of the five species studied produced adventitious roots: F. aurea, C. icaco, M. cerifera, and I. cassine. L. latisiliquum did not exhibit any adventitious sprouting and therefore probably does not possess a genetic tendency for this adaptation in association with flooding.

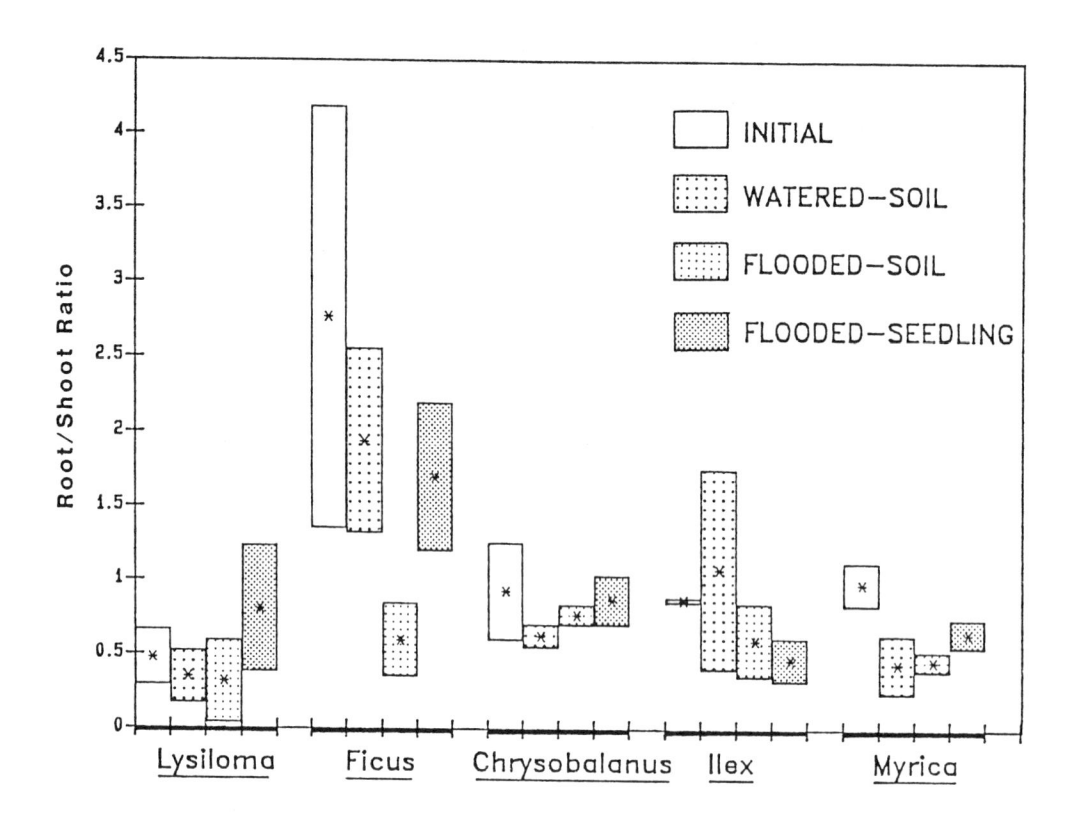

Figure 6. Root:shoot mass ratios for the five tested species in the three flooding treatments at 0 and 90 days. Values are mean root mass per seedling divided by mean shoot mass per seedling.

F. aurea and C. icaco produced roots under a variety of conditions, and the production cannot be specifically attributed to flooding. F. aurea commonly produces aerial adventitious roots in epiphytic terrestrial sites (Tomlinson, 1980), hence the common name "strangler fig". Ficus seedlings in the flooded seedling treatment produced roots that grew above the soil surface in the water to lengths of 30 to 50 cm.

The two species that exhibited a flood root response were M. cerifera and I. cassine, and the degree of response increased with depth of flooding. M. cerifera seedlings in both the flooded-soil and flooded-seedling treatments produced flood roots that originated from the root collar just below the soil surface and grew up into the water. More roots per seedling and longer roots were observed in the flooded-seedling treatment. The completely flooded seedlings of I. cassine exhibited an interesting adaptation by producing roots along the stem at or near the leaf nodes.

Lenticel hypertrophy is a response to stem anoxia. The larger lenticels allow a greater influx of diffused oxygen and also release toxic compounds that are by-products of anaerobosis. (Kozlowski, 1984; Hook and Scholtens, 1978). Numbers of lenticels were counted in this experiment and used to evaluate flood adaptations of the various species. However, numbers of lenticels did not increase in four of the species. <u>F. aurea</u> was the only species with an increase in the number of lenticels over the course of the experiment. The <u>F. aurea</u> lenticel counts increased in all treatments and were not attributed to flooding. Therefore, lenticel numbers do not appear to be a flood adaptation in these species.

<u>Lysiloma</u> was intolerant of minor flooding, based upon little or no height or mass growth in flooded soil, loss of leaves within three days in the flooded-soil treatment, no survival of flooded seedlings, and no morphological response to flooding. These results were expected, as this plant is found in upland or drier sites such as pinelands and hardwood forests (hammocks) in southern Florida, and never in inundated sites.

The remaining species all exhibited tolerance of flooding. Based upon survival in the flooded-seedling treatment, the species can be ranked from least tolerant to most tolerant in the order of <u>Ficus</u>, <u>Chrysobalanus</u>, <u>Ilex</u>, and <u>Myrica</u>. <u>Ilex</u> and <u>Myrica</u> were more flood tolerant than the other two species, based upon statistically significant growth with flooded roots. <u>Myrica</u> also exhibited significant development of adventitious roots, indicative of acclimation to root flooding.

These groupings correspond to the associations of these species found in natural settings. <u>Ilex</u>, <u>Myrica</u>, and <u>Chrysobalanus</u> are common components of swamp forest or tree islands found in the Everglades (Davis, 1943; Craighead, 1971). <u>Myrica cerifera</u> is found along banks or shores of ponds and in swamps throughout the southeastern United States (Godfrey and Wooten, 1982). <u>Ficus</u> is found in bayheads or cypress swamps (Godfrey and Wooten, 1982; Tomlinson, 1980), but generally as an epiphyte on larger cypress trees or growing on raised hammocks around the base of cypress trees. However, it appears that <u>Ficus</u> is tolerant of some flooding, but perhaps not to the extent of the common swamp forest species of <u>Ilex</u>, <u>Chrysobalanus</u>, and <u>Myrica</u>.

ACKNOWLEDGMENTS

This work was supported by the National Park Service, U.S. Department of the Interior. We would like to thank Rosina Hassoun for aiding in early stages of the experiment.

LITERATURE CITED

Alexander, T.R. and A. Crook, 1975. Recent and long-term vegetation changes and patterns in South Florida. Appendix G. Part 2. South Florida Environmental Study. University of Miami, Coral Gables, Florida.

Armstrong, W., 1968. Oxygen diffusion from the roots of woody species. Physiologia Plantarum 21: 539.

Briscoe, C.B., 1957. Diameter growth and effects of flooding on certain bottomland forest trees. Dissertation, Duke University, Durham, North Carolina.

Craighead, F.C., 1971. The Trees of South Florida, Vol. 1: The Natural Environment and Their Succession. University of Miami Press, Coral Gables, Florida.

Davis, J.H., 1943. The Natural Features of South Florida. Florida Geological Survey Bulletin No. 25. Florida Geological Survey, Tallahassee, Florida.

Demaree, D., 1932. Submerging experiments with *Taxodium*. Ecology 13: 258-262.

Gil, C.J., 1970. The Flooding Tolerance of Woody Species-A Review. Forest Abstracts Leading Article Series No. 44, 31: 671-688.

Godfrey, R.K. and J.W. Wooten, 1982. Aquatic and Wetland Plants of the Southeastern United States. University of Georgia Press, Athens, Georgia.

Gunderson, L.H., 1977. Regeneration of cypress, *Taxodium distichum* and *T. ascendens* in logged and burned cypress strands at Corkscrew Swamp Sanctuary. Thesis, University of Florida, Gainesville, Florida.

Hook, D.D. and J.R. Scholtens, 1978. Adaptations and flood tolerance of tree species. In: Plant Life in Anaerobic Environments, D. D. Hook and R.M.M. Crawford (Editors). Ann Arbor Science, Ann Arbor, Michigan. pp. 299-331.

Johnson, L., 1958. A survey of the water resources of Everglades Park. Report Everglades National Park, Homestead, Florida.

Kolipinski, M.C. and A.L. Higer, 1968. Hydrobiological investigations in Everglades National Park. U.S. Geological Survey Open File Report. U.S. Geological Society, Tallahassee, Florida.

Kozlowski, T.T., 1984. Plant responses to flooding of soil. Bioscience 34:162-167.

Kummerow, J., 1980. Adaptation of roots in water-stressed native vegetation. In: Adaptation of Plants to Water and High Temperature Stress, N.C. Turner and P.J. Kramer (Editors). John Wiley and Sons, New York, New York. pp. 57-73.

Loucks, W.L. and R.A. Keen, 1973. Submersion tolerance of selected seedling trees. Journal of Foresty 71:496-497.

Loveless, C.M., 1959. A study of the vegetation in the Florida Everglades. Ecology 41: 1-9.

McPherson, B.F., 1973. Vegetation in relation to water depth in Conservation Area 3A. U.S. Geological Survey Open File Report 73025. U.S. Geological Survey, Tallahassee, Florida.

Putnam, J.A., 1951. Management of bottomland hardwoods. Southern Forest Experiment Station Occassional Paper 116. U.S. Department of Agriculture, New Orleans, Louisiana.

Schwintzer, C.R. and S.A. Lancelle, 1983. Effect of water-table depth on shoot growth, and nodulation of *Myrica gale* seedlings. Journal of Ecology 71: 489-501.

Sena-Gomes, A.R. and T.T. Kozlowski, 1980. Responses of *Melaleuca quinquenervia* seedlings to flooding. Physiologia Plantarum 49: 373-377.

Tomlinson, P.B., 1980. The Biology of Trees Native to Tropical Florida. Harvard University Press, Allston, Massachusetts.

EFFECTS OF REGULATED LAKE LEVELS ON THE AQUATIC ECOSYSTEM OF VOYAGEURS NATIONAL PARK

L. W. Kallemeyn[1], M. H. Reiser[2], D. W. Smith[3], J. M. Thurber[3]

ABSTRACT

A single hydropower facility and two small regulatory dams located outside Voyageurs National Park regulate water levels in the four large lakes that comprise 96% of the park's water area. The present water management program utilizes greater-than-natural fluctuations in water levels on the one reservoir that encompasses three of the lakes to maintain less-than-natural fluctuations on the second reservoir. Additionally, the regulated lake levels differ from natural levels by usually peaking later, remaining relatively stable throughout the summer rather than slowly declining, and on one reservoir by declining 1.8 m rather than 0.6 m over the winter. This regulatory system was found to have an adverse affect on northern pike, common loon, red-necked grebe, beaver, and muskrat populations of the littoral zone and adjacent wetlands of the park. An alternative water management program is presented that would meet the biological requirements of these species by restoring more natural lake-level fluctuations. It is hypothesized that it would also have a positive effect on the other members of the aquatic community.

INTRODUCTION

Voyageurs National Park, which is located along the Minnesota-Canadian border (Figure 1), was established in 1975 to preserve for present and future generations the scenery, natural and historic objects, and wildlife that occur along a portion of the historic fur trade route of the park's namesake, the voyageurs. The park encompasses 88,628 hectares, of which approximately 34,400 hectares or 39% is covered by water. Kabetogama Lake and those portions of Namakan, Sand Point, and Rainy Lakes that lie within the park comprise 96% of the water area. In addition to the four large lakes, there are 26 smaller, inland lakes, as well as numerous small beaver ponds.

Lake levels in the large lakes in Voyageurs National Park have been controlled by a hydroelectric dam at the outlet of Rainy Lake and by two dams at the outlet of Namakan Lake since the early 1900s. The latter two dams control the water levels in Namakan, Kabetogama, and Sand Point Lakes, which collectively are referred to as Namakan Reservoir. Because portions of all the lakes except Kabetogama lie in Canada, the dams and lake levels are regulated

---
[1] Voyageurs National Park, P. O. Box 50, International Falls, Minnesota 56649
[2] Department of Biological Sciences, Northern Arizona University, Box 5640, Flagstaff, Arizona 86011
[3] Department of Biological Sciences, Michigan Technological University, Houghton, Michigan 49931

by the International Joint Commission. The dams and lake levels are managed for the authorized purposes of power generation, navigation, flood control, pollution abatement, and fish propagation.

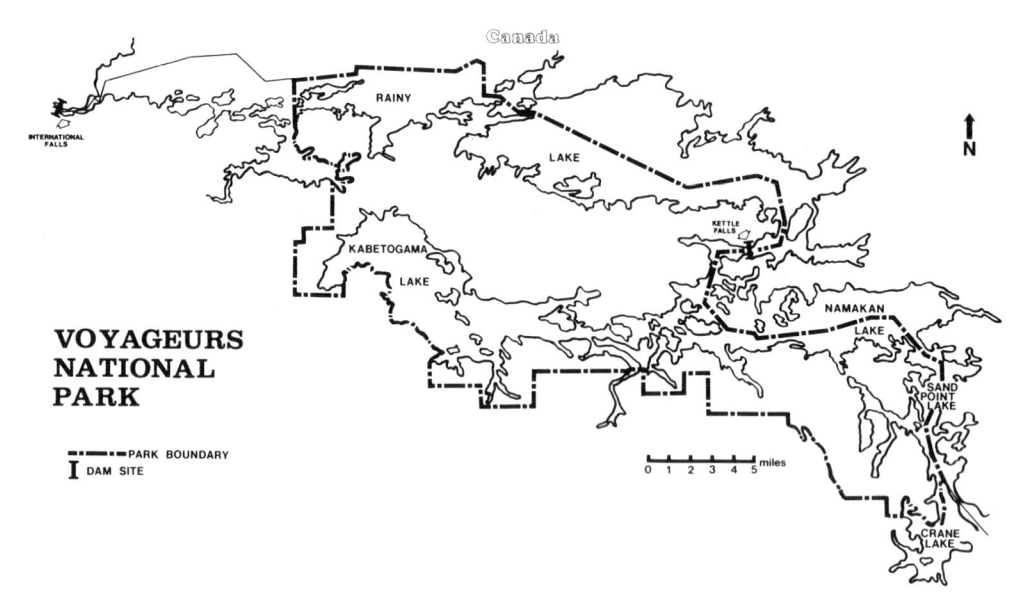

Figure 1. Map of Voyageurs National Park study area.

With the existing water management programs or "rule curves," larger-than-natural fluctuations in lake levels on Namakan Reservoir are used to maintain less than natural fluctuations on Rainy Lake (Figure 2). Namakan Reservoir's average annual water-level fluctuation is about 2.7 m while Rainy Lake's is about 1.1 m. The fluctuation of Namakan Reservoir is about 0.9 m greater than the estimated natural or pre-dam fluctuation while Rainy Lake's is about 0.8 m less (Flug, 1986). The timing of the fluctuations is also different under the regulated system. Regulated lake levels usually peak in late June or early July rather than late May or early June as they did prior to dam construction, remain stable throughout the summer rather than gradually declining, and on the Namakan Reservoir lakes decline 1.8 m over the winter rather than 0.6 m (Cole, 1982).

Concerns about the effects of regulated lake levels on the aquatic biota, particularly the fish community, have existed ever since the dams were constructed (Sharp, 1941; Johnson et al., 1966; Chevalier, 1977). However, the establishment of Voyageurs National Park, with its emphasis on restoring and preserving the natural environment, resulted in a heightened concern about the impacts of the regulated lake levels on the aquatic ecosystem (Cole, 1979, 1982). The impacts on those organisms and plants that occur in the littoral zone, which is the area most affected by the fluctuating water levels, were of particular concern.

Because of these concerns, a research program to assess the impacts of the regulated lake levels on the park's aquatic eco-

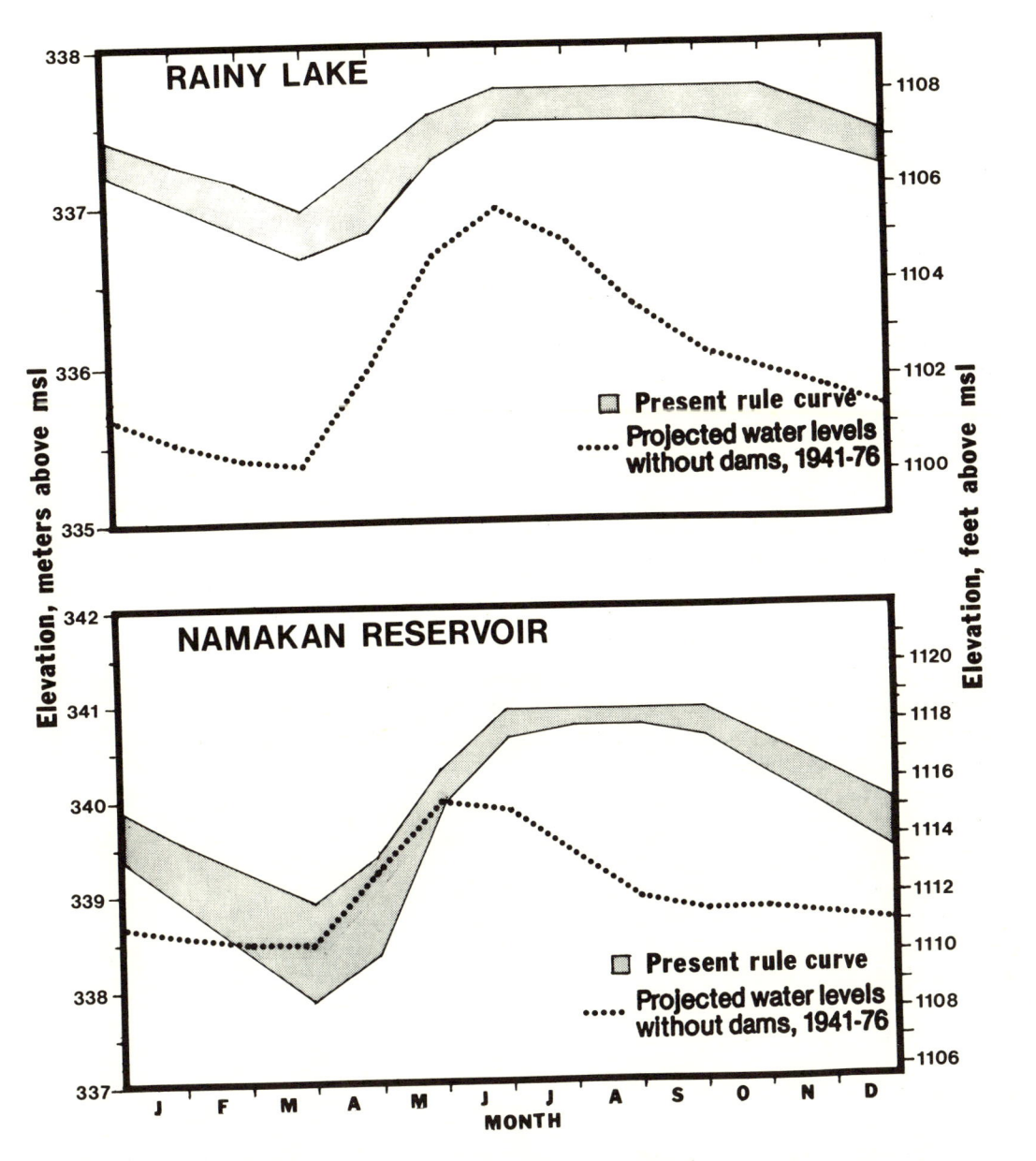

Figure 2. Present water management programs (rule curves) and computed natural lake levels for Rainy Lake and Namakan Reservoir, Minnesota and Ontario. Elevations given in meters or feet above mean sea level (msl).

system and to develop possible alternatives to the present water management program was initiated in 1983 by the National Park Service. The five primary elements in this program were 1) a study of the hydrology of Namakan Reservoir and Rainy Lake and analysis of the impact of fluctuating lake levels on 2) littoral vegetation and benthic organisms, 3) the fish community, particularly northern pike, Esox lucius, 4) shore and marsh nesting birds, particularly the common loon, Gavia immer and red-necked grebe, Podiceps grisegena, and 5) beaver, Castor canadensis, and muskrat, Ondatra zibethicus, colonies. This paper presents results from studies dealing with the latter three elements.

METHODS

Northern Pike

The effects of regulated lake levels on northern pike spawning habitat availability and reproductive success were evaluated in Kabetogama Lake from 1983 through 1985 (Kallemeyn, 1987a). The amount of potential northern pike spawning habitat between 338.9 and 340.5 m above mean sea level (msl) in two tributary deltas in Kabetogama Lake was determined from topographic maps and the results of vegetative cover type surveys. Reproductive success was determined from catches of young-of-the-year in standard seine hauls and trap net sets.

Shore and Marsh Nesting Birds

Studies of common loon and red-necked grebe were conducted from 1983 to 1985 on both the regulated Rainy Lake and Namakan Reservoir and the inland lakes with their naturally-regulated water levels. Intensive shoreline searches were used to determine adult densities and to locate nests of the two species in the park and adjacent waters. Nests and resulting broods were monitored to determine hatching and fledging rates. Nesting habitat, climatological, hydrological, and lake characteristics data were collected for use in evaluating their effects on nesting and reproductive success.

Beaver

The effects of regulated lake levels on winter (December-March) behavior of beaver was primarily determined by locating radio-implanted animals (Telonics Inc., 400 L radio implant with temperature sensitivity). Beavers were located at least twice a week, at which time their position relative to the lodge (in and out), their body temperature, and the time of day were recorded.

Muskrats

The effects of regulated lake levels on muskrat populations were evaluated by comparing densities of muskrat houses in areas of similar habitat from Rainy Lake (0.6 m winter drawdown) and Kabetogama Lake (1.8 m winter drawdown). In 1985, houses were counted in Black and Cranberry Bays in Rainy Lake and Daley Brook in Kabetogama Lake. Tom Cod Creek in Kabetogama Lake was added to the survey in 1986. Total house counts were made by skiing or snowmobiling in January and March in both winters.

RESULTS

Northern Pike

In Kabetogama Lake, over 90% of the emergent vegetation, which is the preferred spawning habitat of northern pike (Fabricius and Gustafson, 1958; Franklin and Smith, 1963; McCarraher and

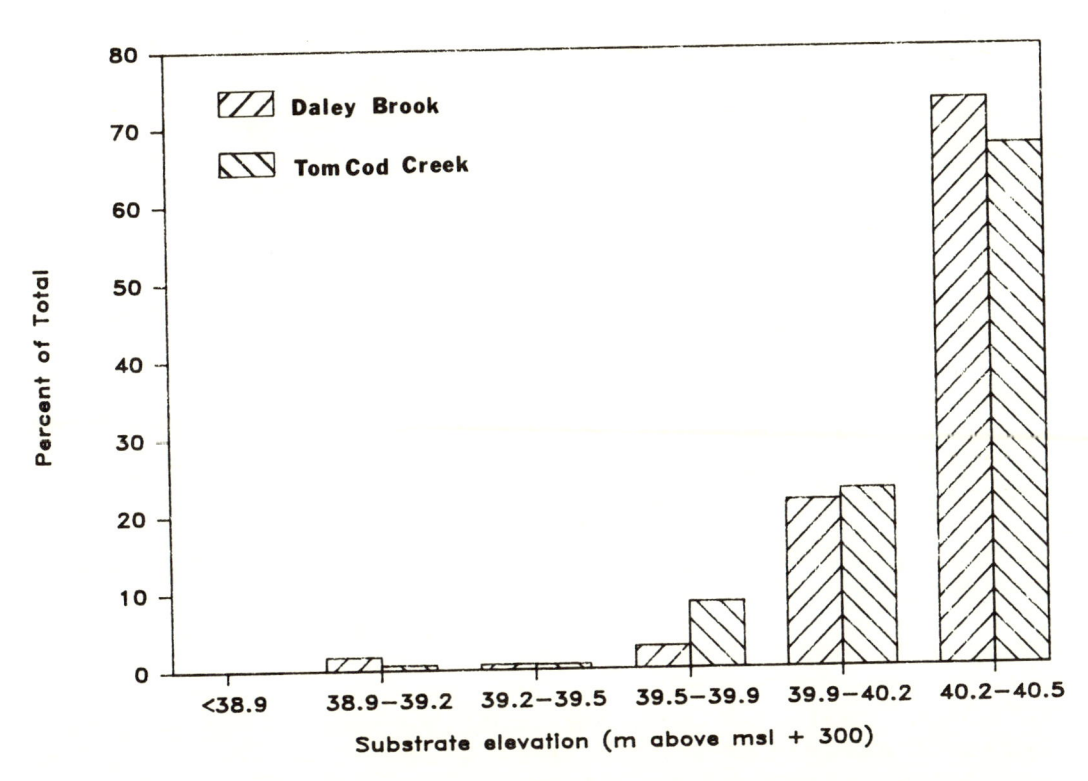

Figure 3. Distribution of emergent vegetation in relation to lake-bed elevation in the deltas of Daley Brook and Tom Cod Creek, Kabetogama Lake.

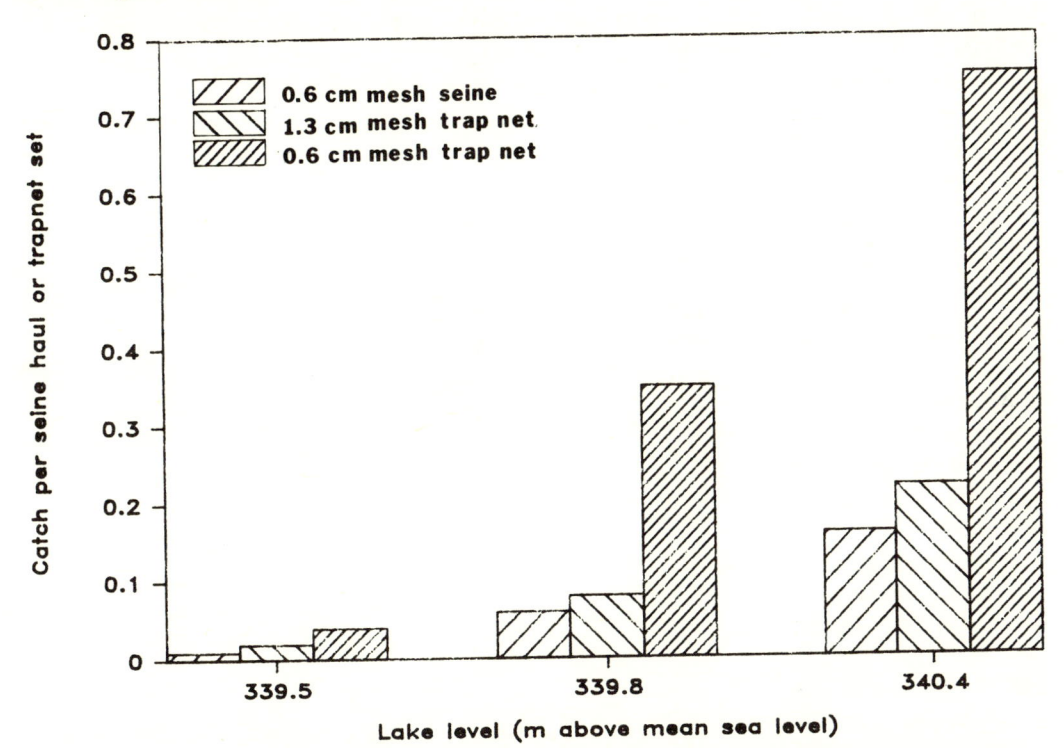

Figure 4. Comparison of catch/seine haul and trap net set (catch/unit effort-c/f) of young-of-the-year northern pike with the water level three weeks after ice-out, Kabetogama Lake, 1983-85.

Thomas, 1972), was found to occur above 339.9 m msl (Figure 3). Consequently, a 1.5 to 1.8 m rise in lake levels is required each spring to flood and provide access to the preferred spawning substrate. When water levels reached the emergent vegetation within three weeks of ice-out, which is when northern pike spawning typically occurs in these lakes, reproductive success was higher (Figure 4). Flooding of the spawning habitat within three weeks of ice-out only occurred, however, because water levels exceeded the maximum levels called for under the current water management program. The water level has exceeded 340.2 m msl within three weeks of ice-out eight out of fifteen years since the present rule curve went into effect in 1971. In all instances where this occurred, however, lake levels were above the upper level of the rule curve.

Shore and Marsh Nesting Birds

Voyageurs National Park's adult loon population of approximately 200 birds produced an average of 27 young per year (Reiser, 1984, 1985). The overall reproductive rate for park loons of 0.34 fledged young per pair of adults per year (yg/pr/yr) was low in comparison to other populations in Minnesota and Saskatchewan (Table 1). This reproductive rate is 30% below the level thought necessary for replacement of individuals in this population (Nilsson 1977). The reproductive rate ranged from 0.22 fledged yg/pr/yr on the Namakan Reservoir lakes to 0.58 fledged yg/pr/yr on Rainy Lake (Table 1). Only on Rainy Lake, with its less-than-natural water level fluctuation, was the loon reproductive rate similar to those of loon populations that nest on naturally-fluctuating lakes (Table 1). The reproductive rate from Namakan Reservoir, with its larger than natural water level fluctuations, was considerably lower than on the lakes that fluctuated naturally.

Table 1. Comparison of loon reproductive rates in Voyageurs National Park, 1983-85, with those from populations from other lakes with naturally regulated water levels.

Location	Fledged young/ pair/year	Reference
Voyageurs National Park		
Parkwide average	0.34	This study
Rainy Lake	0.58	This study
Namakan Reservoir	0.21	This study
Inland lakes	0.26	This study
Responder loons	0.97	This study
Other populations		
East-central Saskatchewan	0.54	Yonge, K. unpub. data
Northcentral Minnesota	0.80	McIntyre, 1975
Northeastern Minnesota	0.56	Mooty and Goodermote, 1985

Reproductive rates for loons on the regulated lakes would have been lower if it were not for "responder" loons, which were loons that changed their nesting sites or chronology of nesting in response to the regulated lake levels. Responder loons had a significantly higher (T-test, P<0.05) reproductive rate of 0.97 fledged yg/pr/yr and contributed an average of 51% of the young on the regulated lakes even though they comprised only 14% of the nesting loons. The higher reproductive rate resulted from the birds either utilizing floating bogs for nesting, renesting, or delaying their initial nesting attempt until water levels had peaked in late June or July. Even with the responder's, 30% of the loon nesting attempts were flooded out each year by rising water levels.

The reproductive success of red-necked grebes was also found to be adversely affected in this regulated lake system. Reproductive rates in both Rainy Lake (0.25 fledged yg/pr/yr) and Kabetogama Lake (0.14 fledged yg/pr/yr) were lower than the 0.56 fledged yg/pr/yr observed by Perry and Block (1983) on some naturally-fluctuating lakes in north-central Minnesota.

While grebe reproductive rates were low on both Kabetogama and Rainy Lakes, higher nest and egg losses were associated with Kabetogama's greater-than-natural water level fluctuations. Nest losses on Kabetogama ranged from 63 to 100%, while egg losses ranged from 60 to 100% from 1983 to 1985. On Rainy Lake, which experiences less-than-natural fluctuations, nest losses ranged from 32 to 69% and egg losses from 20 to 67% during the same period.

The low reproductive rates resulted primarily from red-necked grebe nests being flooded out by rapidly rising water levels in mid-June. A representative chronology of nesting outcomes from 1983 is presented in Figure 5. In 1983, water levels during the period from June 10 to 23 rose at a rate ranging from 1.3 to 2.8 cm/day, which resulted in 39% of all grebe nests being flooded out. Similar patterns were evident in 1984 and 1985.

Beaver

As early as January each winter, beavers in the Namakan Reservoir lakes, which undergo a 1.8 m winter drawdown, spent more time out of their lodges than did beavers in Rainy Lake (0.6 m winter drawdown) or in inland ponds (Figure 6). Due to the large drawdown, the lodges in the Namakan Reservoir lakes were out of the water by January and their entrances were covered by ice sheets. Beavers adapted to these conditions by tunneling into lodges or by building secondary dwellings in air spaces that existed below "hanging ice" along the shore. The latter sites were usually located close to the food cache. Beaver in the La Grande River in Quebec exhibited similar adaptions to fluctuating water levels (Courcelles and Nault, 1984).

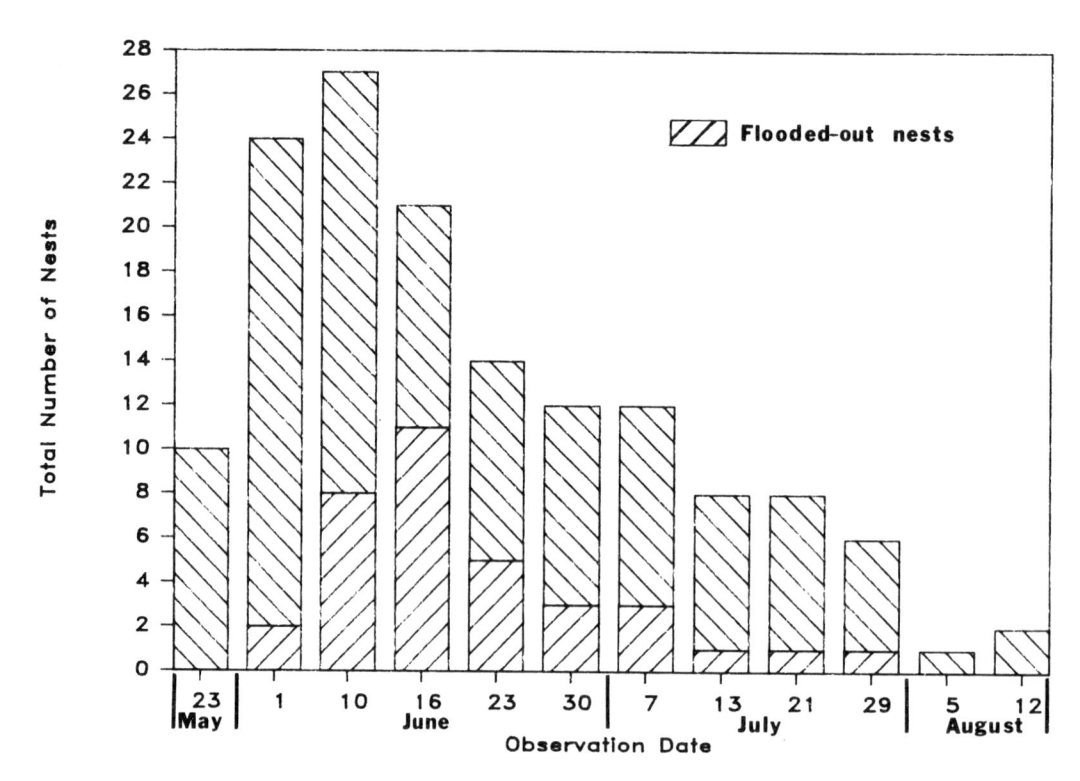

Figure 5. Proportion of active red-necked grebe nests flooded-out throughout the 1983 breeding season in the Rainy Lake study area.

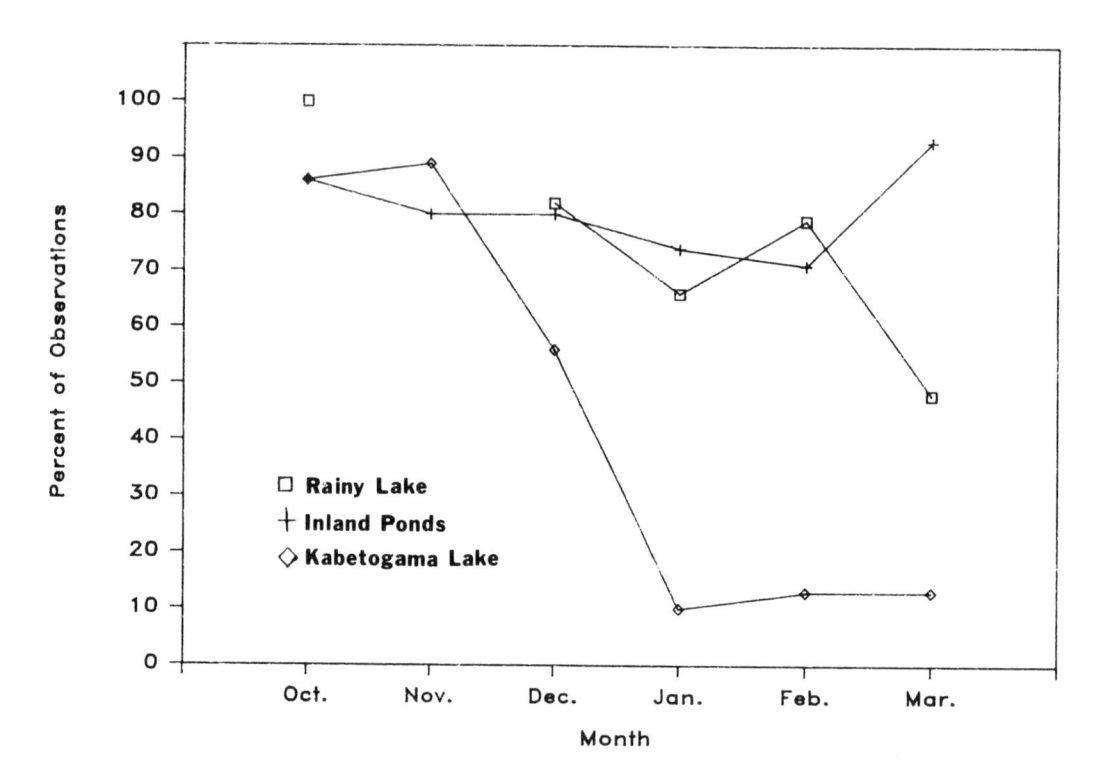

Figure 6. Percent of time radio-tagged beaver were found in lodges during the winter of 1985-86, Voyageurs National Park. No observations were made on Rainy Lake in November.

Beaver in lakes undergoing a 1.8 m drawdown were lighter, lost more weight during the winter, exhibited lower reproductive rates, and moved their homesites more frequently than beaver from the relatively stable inland ponds (Table 2). These differences apparently resulted from the Namakan Reservoir beaver being unable to completely utilize their food caches due to lowered water levels.

Table 2. Comparison of beaver weights (kg) and productivity measures for Namakan Reservoir (drawdown) and inland ponds (non-drawdown), Voyageurs National Park, 1984-85.

	Drawdown n	Drawdown Kg	Non-drawdown n	Non-drawdown Kg
Mean weight (without kits)	51	16.8	48	18.4
Mean weight (kits included)	75	13.1	72	14.3
Mean overwinter weight change	9	-0.2	6	+1.1
Mean kit production/lodge	11	2.18	7	2.29

None of the radio-implanted beaver from either the large or small drawdown lakes died. However, two other beaver that left their lodge in autumn due to the large drawdown subsequently died in March. The move left them without a food cache and both were onshore cutting aspen at the time of their death. One beaver was eaten and presumed killed by a predator while the other animal was found intact but was severely emaciated.

Muskrat

Muskrat house density was significantly different in Rainy and Kabetogama Lakes (T-test, $P<.0005$). At least twice as many muskrat houses were found in bays that experienced a 0.6 m winter drawdown as were found in bays that experienced a 1.8 m drawdown (Table 3). Severe dropping of water levels in winter with consequent freezing-out from food sources has been found to have a detrimental effect on muskrat populations (Errington, 1963). Bellrose and Brown (1941) found twice as many muskrat houses in areas with what they called semi-stable water than in areas with fluctuating water levels in Illinois.

DISCUSSION AND MANAGEMENT RECOMMENDATIONS

All the species investigated were found to be adversely affected by the present water management programs, and in particular by the greater-than-natural water level fluctuations that occur on the Namakan Reservoir lakes. These species have been unable to adjust to the changes in the magnitude and timing of lake-level fluctuations since the dams were constructed, and in particular to the water management program that has been used since 1971.

Table 3. Muskrat house counts for winters of 1984-85 and 1985-86 on Rainy Lake and Namakan Reservoir study areas, Voyageurs National Park.

	Hectares of littoral vegetation surveyed		Houses/ hectare	
	1985	1986	1985	1986
Rainy Lake				
Cranberry Creek	79.6	79.6	0.55	0.44
Black Bay	69.1	15.2	0.59	0.59
Kabetogama Lake				
Daley Brook	57.1	57.1	0.12	0.23
Tom Cod Creek	----	63.2	----	0.21

Implementation of alternative water management strategies that more closely approximate the magnitude and timing of natural fluctuations in lake levels should have a beneficial effect on all these species because they evolved in an ecosystem that included fluctuating water levels. To restore more natural conditions, the water management programs on both Rainy Lake and Namakan Reservoir should be modified so that they provide for higher water levels earlier in the spring, stable levels during June, summer drawdowns of 0.6 and 0.9 m, respectively, and a reduction in the winter drawdown on Namakan Reservoir from 1.8 m to 0.8 m (Figure 7). The latter change will reduce the overall fluctuation on the Namakan Reservoir lakes to 1.7 m (Figure 7).

The proposed changes should have a positive affect on all of the species that were studied. The higher water levels earlier in the spring will benefit northern pike by increasing the frequency with which lake levels on Namakan Reservoir reach the emergent vegetation preferred for spawning during the spawning season. The walleye, another spring spawner, would also benefit from the higher water levels since its reproductive success has also been found to be directly related to lake levels during the two week period following ice-out in Rainy, Kabetogama, and Sand Point Lakes (Chevalier, 1977; Kallemeyn, 1987b). Additionally, the earlier high water levels should have a positive impact on loon and grebe nesting success since they will help meet the goal of maintaining relatively stable water levels during the June nesting period.

The proposed summer drawdowns will also benefit northern pike by providing environmental conditions that will allow emergent vegetation to grow at lower elevations within these lakes. Emergent vegetation will thus be flooded more frequently, even in low runoff years, and will be more readily available for northern pike spawning. Emergent vegetation in freshwater marshes has been found elsewhere to be limited by high, stable water levels, such as those now maintained on Namakan Reservoir throughout the growing season (Harris and Marshall, 1963; Kadlec, 1962). Draw-

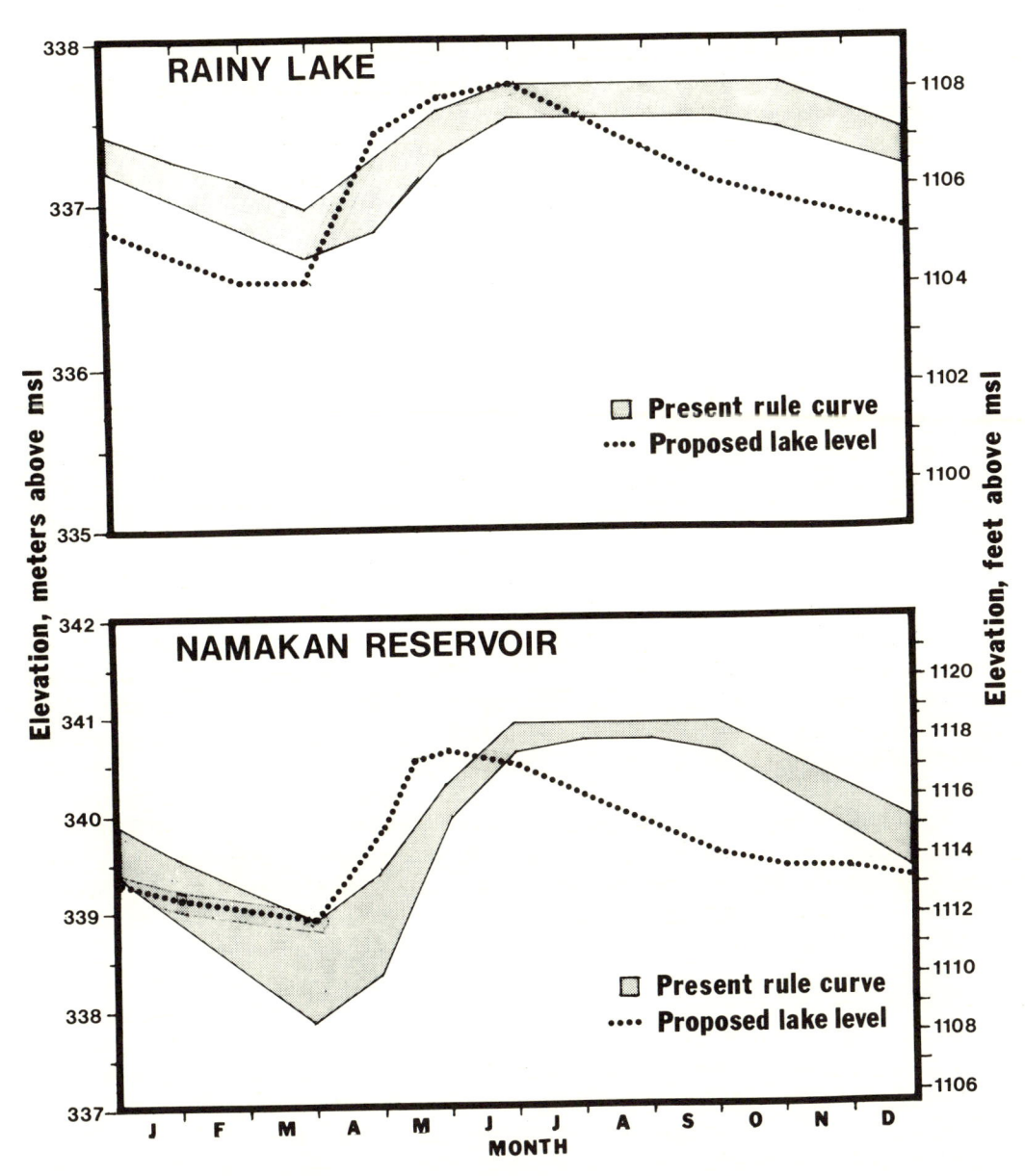

Figure 7. Comparison of recommended alternatives and existing water management plans of "rule curves" for Rainy Lake and Namakan Reservoir, Minnesota and Ontario. Elevations given in meters or feet above mean sea level (msl).

downs are commonly used by wetland and fisheries managers to overcome this limiting effect and to increase the distribution and diversity of emergent plant species (Kadlec, 1962; Groen and Schroeder, 1978; Keddy and Reznicek, 1985). The response of the emergent vegetation to the proposed drawdown could be evaluated using a model developed by van der Valk (1981) for predicting and evaluating the response of various emergent species to drawdowns and other changes in water levels.

Beaver and muskrat in the Namakan Reservoir lakes will also benefit from the summer drawdown, particularly if it is combined with a reduction in the winter drawdown. Presently, high water levels are maintained into early autumn and this causes the animals to establish their food caches and lodges in locations that make them susceptible to dewatering, particularly in those lakes with a 1.8 m winter drawdown. The combination of the summer drawdown and the reduced winter drawdown should overcome these problems and enable the animals to utilize their food caches and lodges throughout the winter. On Rainy Lake, the present 0.6 to 0.9 m winter drawdown seems to have relatively little effect on the beaver and muskrat populations.

Restoration of more natural lake level fluctuations on Rainy Lake and Namakan Reservoir will probably benefit not only these particular species but also the other members of the aquatic community as well. In wetland management, utilization of management procedures that simulate the natural, seasonal, and annual fluctuations in water levels are believed to benefit more plants and animals and to result in a more typical marsh community than artificial management techniques (Ball, 1985; Weller, 1978). Utilization of this management approach is also in keeping with the National Park Service's mandate to protect, perpetuate, and restore natural environments and native species in national parks, such as Voyageurs (Hayden, 1976).

ACKNOWLEDGMENTS

Special thanks must go to G. F. Cole for his continual guidance and support. Thanks must also go to the too-numerous-to-name park employees and volunteers who assisted with these studies. Without their able assistance, much of this work would not have been completed.

LITERATURE CITED

Ball, J. P., 1985. Marsh management by water level manipulation or other natural techniques: a community approach. In: Coastal Wetlands, H. H. Prince and F. M. D'Itri (Editors). Lewis Publishing, Chelsea, Michigan, pp. 263-277.

Bellrose, F. C. and L. G. Brown, 1941. The effect of fluctuating water levels on the muskrat population of the Illinois River valley. Journal of Wildlife Management 5: 206-212.

Chevalier, J. R., 1977. Changes in walleye (<u>Stizostedion vitreum vitreum</u>) population in Rainy Lake and factors in abundance, 1924-75. Journal of the Fisheries Research Board of Canada 34: 1696-1702.

Cole, G. F., 1979. Mission-oriented research in Voyageurs National Park. Proc. Second Conf. on Scientific Research in the National Parks 7: 194-204.

Cole, G. F., 1982. Restoring natural conditions in a boreal forest park. In: Transactions of the 47th North American Wildlife and Natural Resources Conference. Wildlife Management Institute, Washington, D. C., pp. 411-420.

Courcelles, R. and R. Nault, 1984. La Grande River hydroelectric complex: beaver behavior during the exploitation of La Grande and Opinaca hydroelectric reservoirs. Societe d' energie de lu Baie James. 78 pp.

Errington, P. L., 1963. Muskrat populations. Iowa State University Press, Ames, IA. 665 pp.

Fabricius, E. and K. J. Gustafson, 1958. Some new observations on the spawning behavior of the pike, *Esox lucius* L. Fish. Board Sweden, Inst. Freshwater Res., Drottingholm 39:23-54.

Flug, M., 1986. Analysis of lake levels at Voyageurs National Park. Water Resources Div., National Park Service Rept. 86- 5. 52 pp.

Franklin, D. R. and L. L. Smith, Jr, 1963. Early life history of the northern pike, *Esox lucius* L., with special reference to the factors influencing the numerical strength of year classes. Transactions of the American Fisheries Society 92: 91-110.

Groen, C. L. and T. A. Schroeder, 1978. Effects of water level management on walleye and other coolwater fishes in Kansas reservoirs. In: R. L. Kendall (editor). Selected Coolwater Fishes of North America. Am. Fish. Soc. Spec. Publ. 11, pp. 278-283.

Harris, S. W. and W. H. Marshall, 1963. Ecology of water-level manipulations on a northern marsh. Ecology 44: 331-343.

Hayden, P. S., 1976. The status of research on the Snake River cutthroat trout in Grand Teton National Park. In: Research in the Parks. U.S. Department of the Interior, National Park Service Symposium Series, No. 1, pp. 39-47.

Johnson, F. H., R. D. Thomasson and B. Caldwell, 1966. Status of the Rainy Lake walleye fishery, 1965. Minn. Dept. Conserv., Div. Game Fish., Sect. Res. Plan. Invest. Rep. 292, 22 pp.

Kadlec, J. A., 1962. Effects of a drawdown on a waterfowl impoundment. Ecology 43: 267-281.

Kallemeyn, L. W., 1987a. Effects of regulated lake levels on northern pike spawning habitat and reproductive success in Namakan Reservoir, Voyageurs National Park. National Park Service, Midwest Region Research/Resources Management Report (in press).

Kallemeyn, L. W., 1987b. Correlations of regulated lake levels and climatic factors with abundance of young-of-the-year walleye and yellow perch in four lakes in Voyageurs National Park. North American Journal of Fisheries Management (in press).

Keddy, P. A. and A. A. Reznicek, 1985. Vegetation dynamics, buried seeds, and water level fluctuations on the shorelines of the Great lakes. In: Coastal Wetlands, H. H. Prince and F.M. D'Itri (Editors). Lewis Publishing, Chelsea, MI, pp. 33-58.

McCarraher, D. B. and R. E. Thomas, 1972. Ecological significance of vegetation to northern pike, *Esox lucius* L., spawning. Transactions of the American Fisheries Society 101: 560-563.

McIntyre, J. W., 1975. Biology and behavior of the common loon (*Gavia immer*) with reference to its adaptability in a man-altered environment. Ph.D. Dissertation, Univ. of Minnesota, St. Paul.

Mooty, J. J. and D. L. Goodermote, 1985. Common loon numbers in the Knife Lake area-Boundary Waters Canoe Area Wilderness. Loon 57: 12-15.

Nilsson, S. G., 1977. Adult survival rate of the black-throated diver, *Gavia arctica*. Ornis. Scand. 8: 193-195.

Perry, P. and S. Block, 1983. The red-necked grebe with a summary of surveys in Crow Wing Co., Minnesota. Minn. Dept. Nat. Res. Rept. 3 pp.

Reiser, M. H., 1984. Annual report on shoreline and marsh nesting bird study, Voyageurs National Park, Minnesota, 1983. Voyageurs National Park Progress Rept. 34 pp.

Reiser, M. H., 1985. Annual report on shoreline and marsh nesting bird study, Voyageurs National Park, Minnesota, 1984. Voyageurs National Park Progress Rept. 26 pp.

Sharp, R. W., 1941. Report of the investigation of biological conditions of Lakes Kabetogama, Namakan, and Crane as influenced by fluctuating water levels. Minn. Dept. Conserv., Div. Game Fish., Sect. Fish. Res. Invest. Rep. 30: 17 pp + appendicies.

van der Valk, A. G., 1981. Succession in wetlands: a Gleasonian approach. Ecology 62:688-696.

Weller, M. W., 1978. Management of freshwater marshes for wildlife. In: Freshwater Marshes: Ecological Processes and Management Potential, R. E. Good, D. F. Whigham, and R. L. Simpson (Editors). Academic Press, New York, NY, pp. 267-284.

Effects of Simulated Muskrat Grazing on Emergent Vegetation

Thomas R. McCabe[1,2] and Michael L. Wolfe[1]

ABSTRACT

Muskrats (<u>Ondatra zibethicus</u>) are associated with wetlands throughout North America. Their impact on marsh vegetation is well-documented. In recent years, research pertaining to marsh ecosystems has emphasized the relationship of interspersion of marsh vegetation with higher faunal diversity and productivity. Muskrats can provide a natural control of aquatic emergents if managed properly. Experimentation at Fish Springs National Wildlife Refuge, Utah was conducted to evaluate the impact of muskrats on their primary food resource, Olney's threesquare bulrush (<u>Scirpus americanus olneyi</u>). The potential effects of muskrat grazing were determined using exclosures in homogeneous stands of Olney's bulrush. A 4 X 4 randomized block design with varying levels of simulated grazing was employed for monthly replications during the growing season for 2 years. Differences in stem density were significant within clipping rates over time and among rates at each sampling. Differences in vegetative yield, measured as mean dry weight (g/stem/plot), were also significant among the simulated grazing rates. Because muskrats are a major influence on wetland vegetation structure, it is necessary to understand the extent of their impact. The positive effects of muskrats in creating the favorable cover:water ratio of a hemi-marsh can be negated by the continued destructive potential muskrats have on plant regeneration and growth.

INTRODUCTION

The muskrat (<u>Ondatra zibethicus</u>) is an integral part of the dynamics of nearly all aquatic systems in North America. Millions are harvested by professional and sport trappers in nearly every state in the Union (Deems and Pursley, 1978). Because muskrats have great economic, recreational, and ecological value, maintenance of populations that are in balance with their habitat is essential to sound wetland management on public lands.

Muskrat populations are often associated with extensive marshes that are managed by state and federal park or refuge personnel. A responsibility of most wetland managers is to maintain the marshes at their maximum potential for perpetuating wildlife resources. To accomplish this goal, management must encompass a broad-spectrum marsh ecosystem approach. Any management plan oriented toward maintaining a marsh at a level of high productivity and species diversity must take into account the relationship of muskrats to the rest of the marsh ecosystem.

---
[1]Department of Fisheries and Wildlife Sciences, Utah State University, Logan, Utah 84322
[2]Present address: U.S. Fish and Wildlife Service, Alaska Fish and Wildlife Research Center, Fairbanks, Alaska 99701

Recent research on marsh ecosystems has emphasized the role of interspersion of marsh vegetation. Weller and Spatcher (1965) found that maximum bird populations and species diversity occurred when a well-interspersed cover-to-water ratio of 50:50 was reached. They coined the term "hemi-marsh" to describe this condition.

The hemi-marsh repeatedly has been shown to have greater faunal species diversity and productivity than any phase of the marsh cycle (Weller and Fredrickson, 1974; Bishop et al., 1979). Therefore, prolonging this stage is often the goal of marsh management. Muskrats are a dominant force in creating and ultimately destroying the hemi-marsh condition (Weller, 1978). Vegetation eat-outs, "... the destruction of marsh vegetation that results from high populations of muskrats ..." (Sipple, 1979), are recorded in nearly every study of muskrat habitat utilization (Errington, 1963; Sipple, 1979). Proper management of muskrats, in order to maximize wetland diversity potential, necessitates knowledge of its impact on the habitat.

A primary objective of this study was to determine the potential effect of muskrat grazing on the growth and productivity of a dominant, emergent plant species. The experimental design entailed measuring the effects of various simulated grazing rates on the muskrat's primary food resource, Olney's bulrush (Scirpus americanus olneyi). The simulated grazing experiment was conducted within exclosures constructed to prevent actual muskrat grazing activity from interfering with the experiment. Monitoring the effects of simulated grazing provided the basis for assessing the damage and potential recovery capabilities of Olney's bulrush. Field experimentation involved the measurement of 3 vegetative attributes: (1) number of stems grown within an experimental quadrat, (2) mean dry weight per stem within the quadrat, and (3) residual effects of grazing, as measured by the number of stems present in the same quadrats at the beginning of each field season. The combined results of the vegetative data were considered a measure of plant productivity.

STUDY SITE

Research was conducted on the 4000 ha of marsh at Fish Springs National Wildlife Refuge (NWR), Utah. The refuge is located in Juab County on the southwest edge of the Great Salt Lake Desert (Figure 1). Situated on a fault line at the base of the Fish Springs Mountain Range, this isolated marsh is fed by 11 major springs that maintain a constant year-round flow of water. The springs inundate a confined area approximately 48 km^2, within which there are natural depressions and control features that create the expansive marshland. Describing the flora of Fish Springs NWR, Bolen (1964) stated: "If a single species were to be designated as descriptive of the marshes it would be Scirpus [americanus] olneyi."

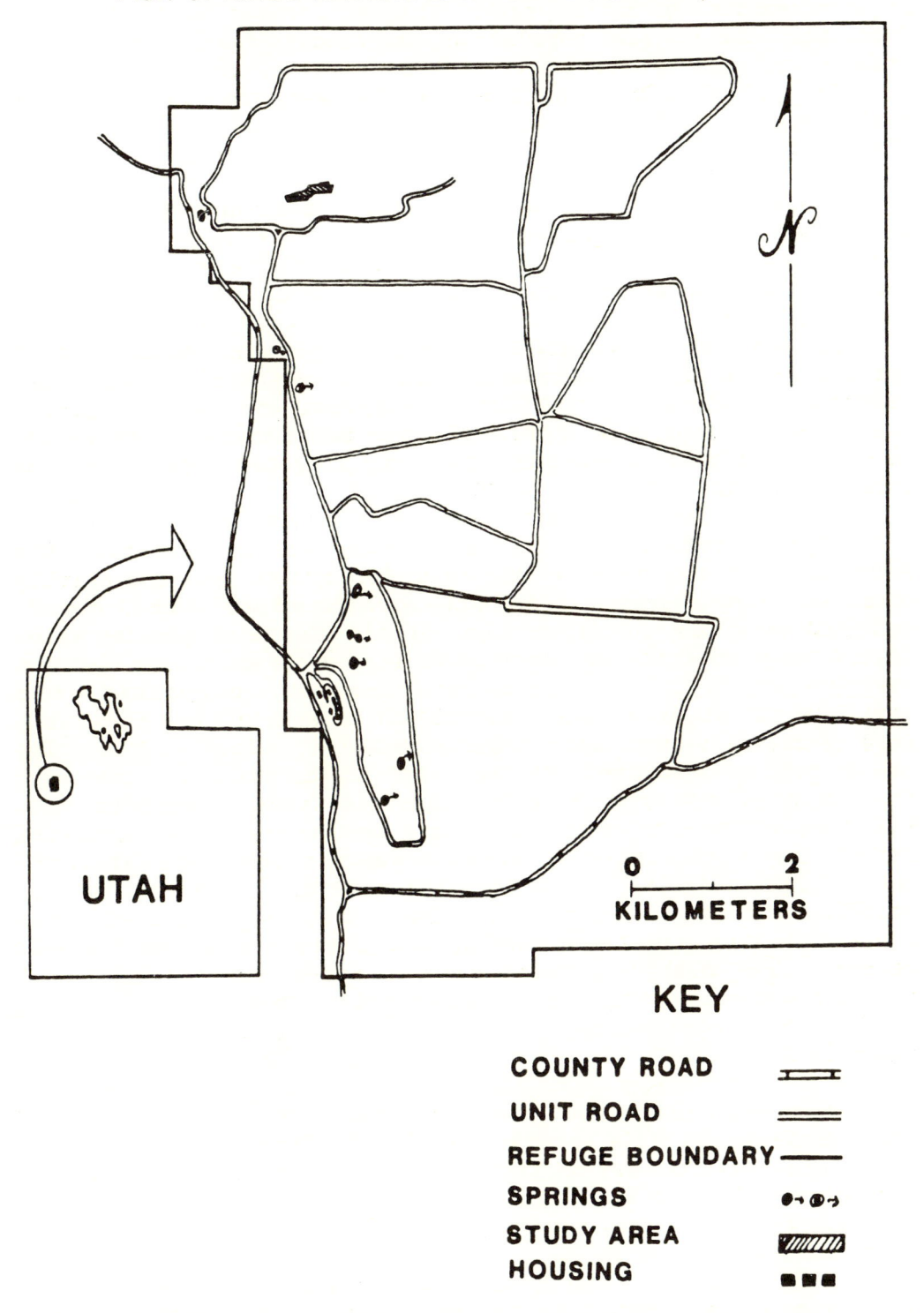

Figure 1. Map of Fish Springs National Wildlife Refuge, Juab County, Utah.

METHODS

Exclosure Site

In July of 1979, an exclosure, 10 X 12 m, was constructed in a homogeneous stand of Olney's bulrush. The exclosure was encircled with 1.0 m high poultry netting, with 0.4 m buried to prevent muskrats from burrowing into the experimental area. The exclosure site was chosen to accommodate 2 annual repetitions of the clipping experiment.

Simulated Grazing Experiment

To test the simulated effects of grazing at different intensities, a 4 X 4 randomized block configuration was utilized. Sixteen plots, 0.5 X 1.0 m, were spaced evenly within the exclosure, leaving a 1.0 m buffer zone between them. Three levels of simulated grazing were tested (33%, 67%, and 100%), with the fourth group used as a control (0% clipped). Initially, all stems in a plot were counted and recorded. Stems were then randomly clipped just above the water line according to the prescribed rate for the plot and in a manner similar to natural grazing by muskrats. Simulations were repeated every 4 weeks. All previously clipped stems that grew to a height $\geq$ 15 cm above the water line between experimental periods were considered to be new stems. During the 1979 field season, 3 sampling dates were possible after the initial July clipping. In 1980, 4 treatments and samplings were conducted between June and September.

In 1980, the experimental plots were moved to minimize confounding effects of the previous year's clipping (Figure 2). Before repositioning the plots, a count was made of stems occurring within the clip-plots utilized during the 1979 field season. These data were used to assess whether grazing created a carry-over impact on the plants' growth. In addition, 16 individual plants of Olney's bulrush were selected and encircled with poultry netting in 1980. These plants were then subjected to the same experimental design as the exclosure areas to ascertain whether individual plants responded differently to the grazing regime. Another experiment, implemented in July 1980, involved differences in plant yield as a determinant of the effects of grazing rates. All stems clipped within the respective plots were dried and weighed separately to determine mean dry weight per stem.

Data Analysis

Data collection within the exclosures involved duplicating the simulated grazing experiment at 4-week intervals. Since the data were collected in the same plots several times during each summer, the samples for a given plot were not independent, and a repeated measures regression analysis was necessary (Draper and Smith, 1981). The mean number of stems for the 4 plots at a given grazing rate was used as the dependent variable because mean

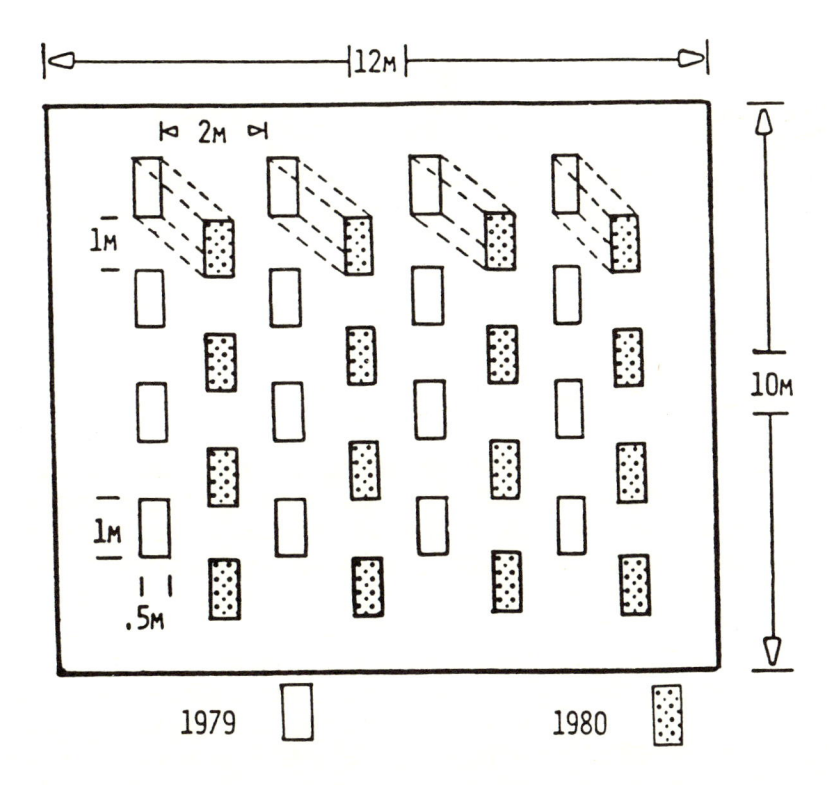

Figure 2. Placement of 0.5 m² quadrats used in the simulated grazing study in Olney's bulrush exclosures for 1979 and 1980.

number of stems per plot was considered the best measure of plant response to the grazing. Months were used as dummy variables in the regression equation (Nie et al., 1975) to form a second degree polynomial. The y-intercepts were assumed to be different for each plot. By testing whether the shapes or the slopes of the curves that were generated were equal, proportional recovery at the various clipping rates could be tested for possible differences. The complete regression model for the data tested the null hypothesis that the number of stems over time within an exclosure for the entire field season was equal (slopes were equal). Since this null hypothesis was ultimately rejected, further partitioning into respective rates of simulated grazing was necessary to test whether the effect of clipping was statistically different among the clipping rates. A split-plot analysis of variance (Steel and Torrie, 1980) was used to assess any difference between the initial number of stems present in the 16 plots in 1979 and those in the same plots at the beginning of the 1980 season (carry-over effect).

Dry weight yield was used as a measure of the relative vigor among the clipped plots (Kershaw, 1973). A two-way analysis of variance of the mean dry weight per stem values was calculated for the 3 clipping rates over 3 months within the exclosure. Since a significant interaction occurred, one-way analysis of variance procedures were used to evaluate the specific effects within rates over time and among rates within a time period.

RESULTS

Number of Stems

Significant differences in numbers of stems did exist among clipping rates. The null hypothesis was rejected because the shapes of the curves generated by the complete model differed significantly (P<0.001) for all exclosure experiments for both 1979 and 1980. The further analyses of 1979 and 1980 data (Table 1) show that the mean plant response to the different clipping levels was significant in all cases. In 1979, 3 replicates of the simulated grazing were conducted from July through September. The curves for Olney's bulrush differed significantly among the rates (P<0.05). Results of the 1980 experiment were comparable, having a probability level of P<0.001. Individual Olney's bulrush plants that were subjected to 1 fewer replication also showed significantly different effects among simulated grazing rates (P=0.014).

Table 1. Results of repeated measures regression analysis of the mean number of stems per 0.5 m^2 quadrat or individual exclosure clipped each month at varying intensities, 1979, 1980.

Data Set	Percent Clipping Rate	June	July	August	September	F	P
Olney's bulrush (1979)	0		242.5	250.0	214.5	3.100	0.043
	33		212.0	165.3	102.5		
	67		230.3	129.3	61.5		
	100		182.3	40.5	14.5		
Olney's bulrush (1980)	0	322.5	377.8	352.8	183.5	9.969	0.000
	33	203.0	225.3	161.0	65.8		
	67	261.0	200.8	121.3	34.5		
	100	207.8	123.3	68.3	7.3		
Individual Olney's bulrush (1980)	0		163.8	180.3	153.8	4.224	0.014
	33		188.5	176.5	105.8		
	67		176.3	115.5	76.5		
	100		205.8	74.8	85.3		

Figures 3 and 4 show the mean number of stems as a plant response to the effects of clipping in 1979 and 1980, respectively. During both experimental seasons, the 0% clipped plots of Olney's bulrush had an initial increase in growth of stems, then a moderate decline in the number of live stems (i.e., aerial green stems) from August through the end of the experiment. Light clipping (33%) produced a similar trend during the first month of 1980 but then a slightly greater loss of stems for the rest of the summer. Both the 67% and 100% (heavy and complete grazing) clipping rates resulted in an immediate decline in the number of

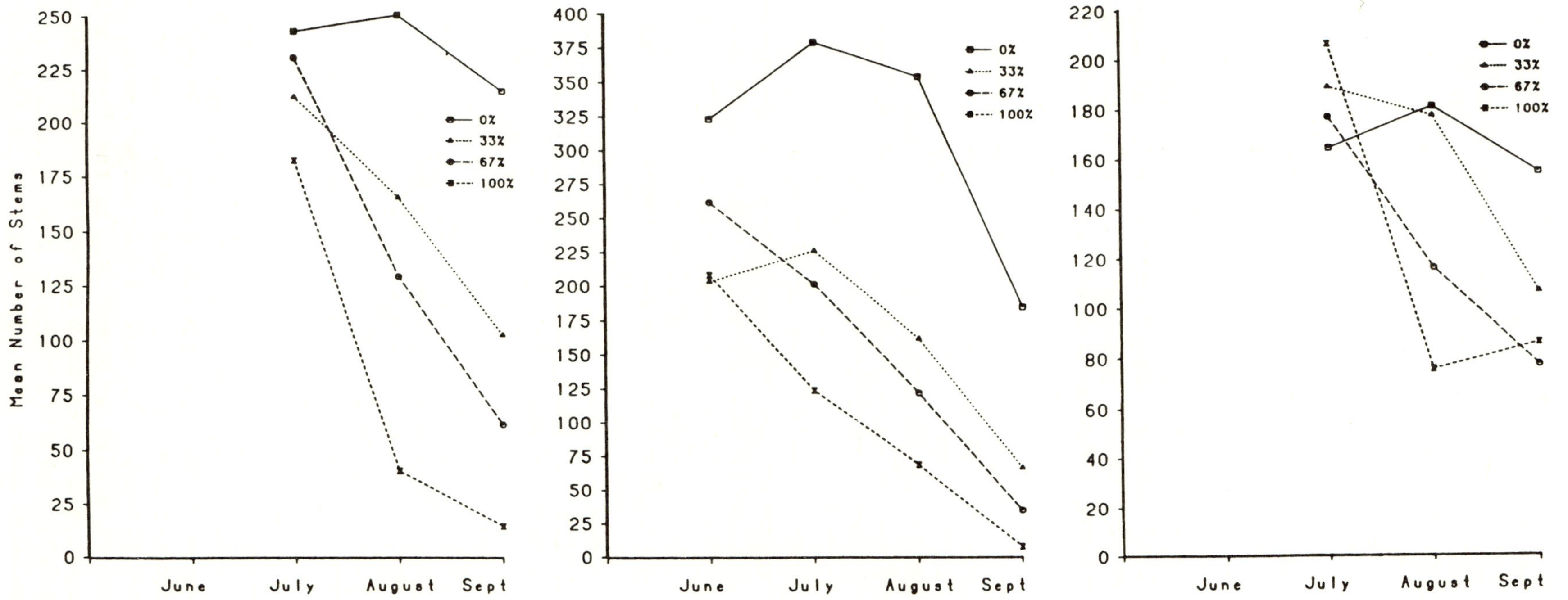

Figures 3, 4, and 5. Mean number of green, aerial stems per simulated grazing rate for each month in the Olney's bulrush exclosures (3) in 1979, (4) in 1980, and (5) individual exclosures, 1980.

stems during both years. The 1979 exclosure and 1980 individually-exclosed plants (Figure 5) of Olney's bulrush were clipped initially in July, at a time when annual growth of the plants is normally waning. Subsequently, the number of stems at all the clipping rates declined.

The split plot ANOVA indicated that there were no significant differences (P=0.453) in numbers of stems among the clipping rates in 1980. The number of stems present in all plots in 1980 was significantly greater (P=0.011) than in 1979. These data (Table 2) suggest that the 1979 clipping did not harm Olney's bulrush in the plots but actually resulted in an overall increase in density.

Table 2. Numbers of Olney's bulrush stems present in the same quadrats at the beginning of the 1979 and 1980 simulated-grazing experiments.

| Year | Mean number of stems/quadrat with std. dev. ||||
	0%	33%	67%	100%
1979	242.3±53.9	212.0±98.2	230.3±140.1	182.3±97.8
1980	275.5±56.1	282.8±30.7	297.3±75.3	253.5±53.4

Dry Weight of Stems

The relative vigor of plants, as measured in dry weight and analyzed by two-way ANOVA, differed significantly among the clipped plots (P=0.000). The one-way ANOVAS (Table 3) showed that differences in mean stem weight among clipping rates within each time period were significant (P<0.03). Differences among months within clipping rates were variable, with light grazing (33%) resulting in a significant increase in weight (P=0.001) and heavy grazing (100%) resulting in a significant decrease (P<0.001), at least from August to September.

Table 3. Combined results of one-way ANOVAs for the effect of clipping rate on mean dry weight per stem of Olney's bulrush within each time period and for the effect within rates over time, 1980.

| Month | Mean dry weight (g)/stem by clipping rate ||| F-value | P-value |
	33%	67%	100%		
July	1.36	1.17	1.00	5.25	0.030
August	1.61	1.27	1.14	9.49	0.006
September	2.16	1.52	0.45	22.69	0.000
F-value	19.53	0.98	36.27		
P-value	0.001	0.411	0.000		

The effects of clipping were not as clearly defined for the individually exclosed Olney's bulrush plants. Plants were initially clipped in July; therefore, the first yield comparisons were not made until August, when growth was near minimum and no differences were observed between clipping rates ($F_{0.05(2,9)}$= 1.33, P=0.31). The mean dry weight per stem decreased linearly with increased clipping rate in September (0.82 to 0.61 to 0.39 g/stem), but the decrease was not significant ($F_{0.05(2,9)}$=3.47, P=0.08).

DISCUSSION

The simulated grazing experiment was designed to be analogous to subjecting the vegetation to different muskrat densities, albeit without root-stock destruction. The intent of the experiments was not only to determine in a general way whether muskrat grazing would impact Olney's bulrush, but specifically to determine whether persistent grazing, a condition occurring at high muskrat densities, affected the plants' ability to recover. Analysis of the exclosure data indicated that the effect of varying intensities of simulated grazing was significant for Olney's bulrush for both years, implying that higher densities of muskrats can potentially affect the productivity of this plant species by more intensive grazing pressure.

The clipping rate always had a significant effect on the number of stems and the mean dry weight per stem. In fact, the effect was more pronounced over time at higher clipping rates, clearly reflecting the level of clipping that Olney's bulrush could tolerate without adverse effects. At light (33%) grazing rates, growth of bulrush seemed to be stimulated. At the 67% grazing rate, stems never adequately overcame the effects of cutting, and at the 100% grazing rate, periodic clipping reduced stem growth to near zero by September.

Ranwell (1961) noted that non-destructive grazing increased the stem density of *Spartina* in salt marshes by up to an order of magnitude. Reimold et al. (1975), during a simulated grazing study, also found that salt marsh vegetation is stimulated to grow when moderately grazed. They recorded increases in both stem density and mean dry weight biomass over that of control plots. While discussing the potential of herbivore grazing at different intensities Odum (1971) stated that "... the light intensity use in which about a third of the net production is removed by grazing, is the optimum use." Allan (1956) found that light cattle grazing enhanced Olney's bulrush production but that the plants were destroyed when over-grazed. Chabreck (1968) noted similar effects during his study of cattle grazing in a Louisiana marshland.

Clipping of plots in 1979 resulted in greater numbers of stems in 1980, so grazing does not seem to have a lasting negative effect on plant density. A possible cause for the increased number of stems is the increased amount of light, and hence warmer sediment temperatures, that would result from less dead plant material from the previous year's growth. One might infer from the data in Table 2 and the results of other researchers that grazing Olney's bulrush stems at all levels of intensity could have a short-term stimulatory effect on growth of the plants. In fact, this would not be the case, particularly at higher grazing intensities. If the plants were heavily grazed, their ability to produce would decrease substantially, and the muskrats would begin to consume the shallow-rooted rhizomes (Lee et al., 1976). This, in turn, would create an eat-out by eliminating the plants altogether. The potential long-term effects of heavy grazing would be deterioration of marsh habitat by creation of large open bodies of water (Roby, 1974) or invasion of less-desirable plant species to the exclusion of Olney's bulrush.

Olney's bulrush will flower from June through September (Tatnall, 1946), with the majority of the seeds being set by early July (Palmisano, 1967). Both Allan and Anderson (1955) and Lay (1945) found this species to grow best during the cooler spring months but very little in hot summers. The Olney's bulrush observed at Fish Springs NWR appeared to grow through mid-July; then above-ground growth diminished. The different responses to clipping between 1979 (June clipping) and 1980 (July clipping), as shown in Figures 3 and 4, demonstrate that the recovery from grazing is affected by seasonality. This could have implications in marsh management.

Olney's bulrush has been considered by different researchers as both a subclimax (Hoffpaur, 1961; Joanen and Glasgow, 1965; Perkins, 1968) and a disclimax (Penfound, 1952) vegetation type, the former preceding climax species, the latter following perterbation of the climax species (Oosting, 1956). Both conditions can be maintained by extrinsic factors (i.e. fire, mowing, grazing). When Olney's bulrush is eliminated, other species will often replace it. At Fish Springs NWR, Bolen (1964) asserted that Olney's bulrush was a climax species forming closed, homogeneous stands. Nelson (1954) also characterized Olney's bulrush in Utah as forming dense, closed stands that eliminate competition from other vegetation species. According to Lay and O'Neil (1942), once the stands are opened, Olney's bulrush can be outcompeted by other species, such as salt-meadow grass (<u>Spartina patens</u>). Our observations at Fish Springs indicated that after a muskrat eat-out, alkali bulrush (<u>Scirpus maritimus</u>) invaded.

The philosophy of multiple use and integrated management has been adopted in principle by state and federal agencies. Therefore, the concept of single-species management cannot be considered to be in the public interest. The role of the muskrat in creating the hemi-marsh condition is well-documented. Equally well-known are the effects on marsh vegetation when muskrats are not regulated properly. Consequently, muskrat population management should be assimilated within broader-spectrum marsh ecosystem management objectives. If compatible with federal and state management goals, marsh managment should be directed toward maintaining the wetland in its most natural, diverse, and productive state. An understanding of the dynamic forces, such as the impacts of muskrat grazing, that normally affect marshes, and, subsequently, their productivity is necessary to formulate appropriate management decisions, particularly on parks and refuges that are maintained for the welfare of wildlife and for the benefit of the general public.

LITERATURE CITED

Allan, P. F., 1956. A system for evaluating coastal marshes as duck winter range. Journal of Wildlife Management 20:247-252.

Allan, P. F. and W. L. Anderson, 1955. More wildlife from our marshes and wetlands. In: The Yearbook of Agriculture. U.S. Department of the Interior Publication, pp. 589-596.

Bishop, R. A., R. D. Andrews, and R. J. Bridges, 1979. Marsh management and its relation to vegetation, waterfowl, and muskrats. Proceedings of the Iowa Academy of Sciences 96:50-56.

Bolen, E. G., 1964. Plant ecology of spring-fed marshes in western Utah. Ecological Monographs 34:143-166.

Chabreck, R. H., 1968. The relation of cattle and cattle grazing to marsh wildlife and plants in Louisiana. Proceedings of the Southeastern Assoc. of Game and Fish Commissioners 22:55-58.

Deems, E. F., Jr. and D. Pursley, 1978. North American Furbearers, Their Management, Research and Harvest Status in 1976. Maryland University Press, College Park.

Draper, N. R. and H. Smith, 1981. Applied Regression Analysis. Wiley and Sons, New York.

Errington, P. L., 1963. Muskrat Populations. Iowa State University Press, Ames.

Hoffpaur, C. M., 1961. Methods of measuring and determining the effects of marsh fires. Proceedings of the Southeastern Association of Game and Fish Commissioners 15:55-58.

Joanen, T. and L. L. Glasgow, 1965. Factors influencing the establishment of wigeon grass stands in Louisiana. Proceedings of the Southeastern Association of Game and Fish Commissioners 19:78-92.

Kershaw, K. A., 1973. Quantitative and Dynamic Plant Ecology. American Elsevier Publishing Company, New York.

Lay, D. W., 1945. Muskrat investigations in Texas. Journal of Wildlife Management 9:56-76.

Lay, D. W. and T. O'Neil, 1942. Muskrats on the Texas coast. Journal of Wildlife Management 6:301-311.

Lee, C. R., R. E. Hoeppel, P. G. Hunt, and C. A. Carlson, 1976. Feasibility of the functional use of vegetation to filter, dewater, and remove contaminants from dredged material. U.S. Corps of Engineers Dredge Material Research Program Technical Report D-76-4, 34 pp.

Nelson, N. F., 1954. Factors in the development and restoration of waterfowl habitat at Ogden Bay Refuge, Weber County, Utah. Utah Department of Fish and Game Publication Number 6, 87 pp.

Nie, N. H., S. H. Hull, J. G. Jenkins, K. Steinbrenner, and D. H. Bent, 1975. Statistical Package for the Social Sciences (SPSS). McGraw-Hill Book Co., New York.

Odum, E. P., 1971. Fundamentals of Ecology. W. B. Saunders, Philadelphia.

Oosting, H. J., 1956. The Study of Plant Communities. W. H. Freeman and Company, San Francisco.

Palmisano, A. W., Jr., 1967. Ecology of *Scirpus olneyi* and *Scirpus robustus* in Louisiana coastal marshes. M. S. Thesis. Louisiana State University, Baton Rouge, Louisiana, 145 pp.

Penfound, W. T., 1952. Southern swamps and marshes. Botanical Review 18:413-446.

Perkins, C. J., 1968. Controlled burning in the management of muskrats and waterfowl in Louisiana coastal marshes. Annual Proceedings of the Tall Timbers Fire Ecology Conference 8:269-280.

Ranwell, D.S., 1961. *Spartina* salt marshes in southern England. I. The effects of sheep grazing at the upper limits of *Spartina* marsh in Bridgewater Bay. Journal of Ecology 49:325-340.

Reimold, R. J., R. A. Linthurst, and P. L. Wolf, 1975. Effects of grazing on a salt marsh. Biological Conservation 8:105-125.

Roby, D., 1974. Marsh regression at the Blackwater National Wildife Refuge. 16 pp. Mimeograph.

Sipple, W. S., 1979. A review of the biology, ecology, and management of *Scirpus olneyi*. Volume II: A synthesis of selected references. Maryland Department of Natural Resources Water Resources Administrative Publication Number 4, 85 pp.

Steel, R. G. D. and J. H. Torrie, 1980. Principles and Procedures of Statistics. McGraw-Hill Company, New York.

Tatnall, R. R., 1946. Flora of Delaware and the Eastern Shore. Society of Natural History of Delaware, Dover.

Weller, M. W., 1978. Management of freshwater marshes for wildlife. In: Freshwater Wetlands: Ecological Processes and Management Potential. R. E. Good, D. F. Whigham and R. L. Simpson (Editors). Academic Press, New York, pp. 267-284.

Weller, M. W. and L. H. Fredrickson, 1974. Avian ecology of a managed glacial marsh. Living Bird 12:269-291.

Weller, M. W. and C. E. Spatcher, 1965. Role of habitat in the distribution and abundance of marsh birds. Iowa Agriculture and Home Economics Experiment Station Report Number 43.

INDEX

-A-

Acer rubrum 2-3
Acid deposition 72, 78-79
Acid lake treatment 70
Acid precipitation 69, 76-77
Adventitious roots 119-120, 122, 126, 129-130
Aerial photography 115
Alkalinity 39, 61, 73-74, 76
Aluminum 57, 60-64
Anaerobic conditions 105
Anaerobosis 128-130
Animal ecology 1, 4, 6-7
Animals 99-101, 104
Anthropogenic degradation 76
Aquifer 37, 40, 43, 45, 49
Artesian 6

-B-

Beaver 133, 135-136, 140-141, 143-144
Bedrock 98-99, 104, 108, 116
Benthos 101, 104
Bicarbonate-sulfate type waters 47
Bicarbonate type waters 45
Bidens cernua 85
Big Cypress Swamp 97, 99, 114-115
Biomass 101, 104
Blag Slough 58-59, 63
Bogs, plant succession in 13-18
Boron 57, 60-64
Buffering systems 70, 74

-C-

C-14 dating 104, 110, 116
Calamagrostis 2-3
Calcium 57, 60-61
Calumet beach 25-27, 29-33
Calumet Lacustrine Plain 39
Canada 133-134
Cape Cod, MA 69-71, 77
Cape Cod National Seashore 69-80
Carbon dioxide 76

Carex 2-3, 87
Castor canadensis 135
Cattail, invasion by 18-19
Chara 88
Charcoal 72, 74, 76-77
Charcoal analysis 11, 18-19, 21
Chenopodium rubrum 85, 88
Chrysobalanus icaco 119-123, 125-127, 129-131
Climate 97-98, 100, 102, 114
Clipping rate 150-152, 154-155
Coal 57-58
Common loon 133, 135-136, 138-139, 142
Cores 43
Corkscrew Swamp 97-118
Correlation analyses 72, 74
Cover:water ratio 147-148
Cowles Bog 1-9, 11, 15-18, 26, 28, 31, 34, 49-50
Cowles, Henry Chandler 2
Cyperus odoratus 88
Cypress 98-99, 104, 106, 108-114

-D-

Dams 133-134, 141
    hydroelectric 133
    regulatory 133
Decomposition 101, 104
Deposits, sedimentary
    calcareous clay 27-29, 31-33
    marl 28-31, 33-34
    peat 28-31, 33-34
    sand 27, 29-31, 33
    till 30-31
Dewatering 4-5
Diatoms 72, 74, 77
Distichlis spicata 85
Drainage 97-98, 113-114
Drought 98-99, 110, 114
Dry weight yield 147-148, 150-151, 154-155
Duck Pond 69, 71, 76
Dune-beach complexes 40
Dunes Creek 26-27, 33
Dune-wetland margin 47

-E-

Electrical Power Research Institute 69, 77-78
Elemental composition 104
Emergent vegetation 136-138, 142-143
Environmental gradients 90
Environmental problems 69, 78
Environments, sedimentary
    coastal 25
    eolian 27, 29, 31
    lacustrine 27-31
    palustrine 27-28, 32
Esox lucius 135
Evaporation 101, 103
Evapotranspiration 101, 103
Everglades National Park 119-132
Exclosure 148, 150-153, 155
Exotics 97

-F-

Farming 98
Fen 2, 4-5
Ficus aurea 119, 121-124, 126-127, 129-131
Fire 97, 99-101, 105, 110, 112, 114-115
Fire in peatlands 18
Fish 101, 104
Fish Springs National Wildlife Refuge 147-158
Flood tolerance 119-132
Flooding 4, 114
Flowways 98
Fly ash 4-6, 57-67
Forest fires 76
Fossil pollen 6, 11-23, 64
Freezes 98, 100, 102-103
Freshwater kettle ponds 69-71, 77-78

-G-

Gavia immer 135
Geohydrology 37-55
Geology 97
Grazing 98, 101
Great Marsh 25-34, 37-39, 43, 45, 49, 52
Great Ponds 69, 77-78

Ground water 2, 4-6, 34, 37-55, 57-59, 63, 74, 77, 100, 103, 106, 108

-H-

Hardwoods 99, 104, 106, 108, 110, 119-120, 131
Hemi-marsh 147-148, 156
Holocene 76
Hurricanes 98, 100, 102
Hydrochemical data 39
Hydrochemistry 37-55
Hydrology 1, 5-7, 63
Hydroperiod 103, 105-107, 113-115
Hypsithermal 32

-I-

Ilex cassine 119, 121-123, 125-127, 129-131
Impoundment 114
Indiana Dunes National Lakeshore 1-9, 11-23, 25-36, 37-55, 57-67
Industrial development 2
Interdisciplinary 1-9, 52, 57, 63, 77-78
Interdunal wetlands 40
International Joint Commission 134
Interspersion of marsh vegetation 147-148
Intradunal wetlands 40, 57
Iron 57, 60-62

-K-

Kabetogama Lake 133, 136-137, 140-142

-L-

Lake Michigan 2, 6, 25, 27, 32-34, 37-40
    changes in level 15, 18
Lake Nipissing 25, 32
Lakeshores 84, 87
Land-use policies 78
Larix laricina 2-3
Leaf abscission 128
Leaf chlorosis 128

Leaf epinasty 128
Lenticel hypertrophy 121, 130
Life-history strategies 89
Liming 69, 76-77
Litter 97, 99-101, 104, 110
Littoral zone 86
Living Lakes 69, 77-78
Logging 98, 101, 113
Lysiloma latisiliquum 119, 121-124, 126-129, 131

-M-

Magnolia virginiana 119
Manganese 57, 60-64
Marl 98-99, 108, 110, 116
Marsh 98-99, 104, 106-109, 112, 116, 147-148, 156-157
Miller Woods ponds 11, 18-21
Minnesota 133
Model 4-5, 99-102
Molybdenum 57, 60-62
Muskrat 133, 135-136, 141-144, 147-148, 150, 155-156
Muskrat eat-out 148, 156
Muskrat grazing 147-148, 150, 155-156
Muskrat management 148, 156

-N-

Namakan Lake 133-134
Namakan Reservoir 133-136, 138-139, 141-144
Naturally-acid ponds 69-70, 72, 74, 76
    biological communities 69
    ecology 69, 74, 76
    management 69-70, 77-78
    paleoecology 69, 74, 76
Nickel 57, 60-63
Nitrates 72
Northern pike 133, 135-137, 142
Nuclear power plant 4, 6
Nutrients 97, 100, 103

-O-

Off-road vehicles 98
Olney's bulrush 147-158
Ondatra zibethicus 135
Outwash plain, pitted 70-72
Overstory vegetation 100

-P-

Paleoecology 1, 7, 69
Peat 98-100, 106-111, 115-116
Peatland 2
Peat mound 4-6, 49
Percent cover 104
Periphyton 99
Persea borbonia 119
pH 57, 60-62
pH-diatoms 69, 72, 74-75, 77
pH-histories, long term lake 70, 74, 76-77
pH-measurements 72-73
    ponds 72, 76
    rain 72
pH-reconstruction 69-70, 72, 74, 76
Phragmites communis 2-3, 87
Phytotoxic 57, 63-64
Pine, pitch 76
Pine, white 76
Pinelands 98-99, 104, 106, 108
Pinhook Bog 11, 13-15
Plant ecology 1-2, 6-7
Plant regeneration 147
Plant succession 11, 19
Plant uptake 63-64
Podiceps grisegena 135
Pollen 72, 74, 77, 101, 104, 115-116
Pollen analysis 11-21
Polygonum lapathifolium 85, 88
Polypogon monspeliensis 85
Ponds 99, 104, 108, 112
Pore waters in peat 49
Potassium 57, 60-61
Prairie marshes 84, 86
Precipitation 69, 77, 98, 100-101, 103, 106
Processes 25-26
Productivity 101, 104, 106, 108

-R-

Rainy Lake 133-136, 138-144
Randomized block design 147, 150
Ranunculus sceleratus 85, 88
Red-necked grebe 133, 135-136, 139-140, 142
Regulated lake levels 133-134, 136

Relative humidity 101
Remote sensing 1, 6-7
Residual effect 148
Root metabolism 129
Root/shoot ratio 129
Roots 99, 102, 104, 112
Rorippa islandica 85, 88
Rumex crispus 85
Rumex maritimus 88

-S-

Salix caroliniana 119, 128
Sand 98, 108-109
Sand Point Lake 133-134
Scirpus acutus 87
Scirpus americana olneyi 147-148
Scirpus fluviatilis 87
Scirpus maritimus 88
Scirpus validus 84-85, 87-88
Scolochloa festucacea 85
Seasonality 72
Seawater ions 72, 74
Secondary succession 57, 63
Sediment analysis, loss on ignition 12-13, 15-16, 18
Sediment cores 71-72, 74-75, 115-116
Sedimentology 1, 6-7, 25-36
Seed bank 64, 81-95
    development 84
    sampling 83
    seedling assay 87
    seedling emergence method 82-83
    seedling identification 82
    sieving seeds 82
    statistical analysis 84
    studies 81-82
Seed dispersal 83-84, 86-87
Seed germination 82-83
Seed longevity 82, 87, 89
Seed production 84
Seed rain 81
Seeds 99
Seepage 4-5, 57-58, 64
Settlement, European 74-76
Settling pond 4-5, 57-59, 61
Shell 99, 105-106, 108
Shrubs 104, 108
Simulated grazing 147-148, 150-155

Site-specific research 69-70, 78
Soil chemistry 1, 5, 7, 57, 64, 101
Soil salinity 86
Soils 97-101, 103, 105-106, 108-110, 112
Soils mapping 64
Sparganium spp. 87
Species composition 101, 106, 108
Specific conductance 74
Sphagnum bog 13
Spring mire 2
Stem numbers 147-148, 150-155
Stimulatory effect 156
Stratigraphy 1, 4-7, 25-36
Strontium 57, 60-62
Succession 89
Successional patterns 99, 101-102, 104, 114-115
Sulfate 57, 61, 72
Surface samples 75, 77
Surface water 4, 57-59, 98, 100, 103
Swamp forest 15, 18

-T-

Temperature 100-101, 103
Test holes 43
Thuja occidentalis 2-3
Tidal influences, ponds 72
Toleston beach 25-27, 29-30, 33
Topography 97, 103, 109
Toxic deposition 72, 78-79
Transfer function 74
Tree rings 101, 104, 113, 115
Typha 2-3, 82, 85, 87-88

-U-

Upward leakage 49

-V-

Vegetation 70, 76
Vegetation dynamics 89
Vegetation recolonization 84
Vegetation zonation 89
Vegetation map 64
Vegetation understory 100, 104

Vibracores 25-28, 30-31, 34
Voyageurs National Park 133-146

-W-

Water budget 102-103
Water chemistry 1, 4-7, 37-55, 57-67, 70, 72, 76-77
Water flows 101, 103
Water level 98-103, 105-108, 111, 113-114
Water level changes 4-6
Water level manipulations 90
Water management 119-120, 133-135, 138, 141-143, 147, 156
Water quality 101, 103
Water table 98, 105-108, 110
Water table profiles 47
Watershed 98, 103, 106
Wetland classification 89
Wetland development 11-23, 25-36
Wetland, freshwater tidal 84
Wetland, lacustrine 85
Wetland management 90, 100, 102, 116, 140, 147, 156-157
Wetland, reclaimed 90
Wetland restoration 90
Wildlife management 91
Woodfall 104

-Z-

Zannichellia palustris 87-88
Zinc 57, 60-64